A-LEVEL YEAR 2

STUDENT GUIDE

EDEXCEL

Chemistry

Topics 16–19

Kinetics II

Organic chemistry II

Organic chemistry III

Modern analytical techniques II

George Facer

HODDER
EDUCATION
AN HACHETTE UK COMPANY

Hodder Education, an Hachette UK company, Blenheim Court, George Street, Banbury, Oxfordshire OX16 5BH

Orders

Bookpoint Ltd, 130 Park Drive, Milton Park, Abingdon, Oxfordshire OX14 4SE

tel: 01235 827827

fax: 01235 400401

e-mail: education@bookpoint.co.uk

Lines are open 9.00 a.m.–5.00 p.m., Monday to Saturday, with a 24-hour message answering service. You can also order through the Hodder Education website: www.hoddereducation. co.uk

© George Facer 2016

ISBN 978-1-4718-5844-4

First printed 2016

Impression number 5 4 3 2 1

Year 2020 2019 2018 2017 2016

This guide has been written specifically to support students preparing for the Edexcel A-level Chemistry examinations. The content has been neither approved nor endorsed by Edexcel and remains the sole responsibility of the author.

Cover photo: TT studio/Fotolia

Typeset by Integra Software Services Pvt. Ltd, Pondicherry, India

Printed in Slovenia

Hachette UK's policy is to use papers that are natural, renewable and recyclable products and made from wood grown in sustainable forests. The logging and manufacturing processes are expected to conform to the environmental regulations of the country of origin.

Contents

Content Guidance

Questions & Answers

■ Getting the most from this book

Sample student answers

Practise the questions, then look at the student answers that follow.

Commentary on sample student answers

Read the comments (preceded by the icon **e**) showing how many marks each answer would be awarded in the exam and exactly where marks are gained or lost.

Commentary on the questions

Tips on what you need to do to gain full marks, indicated by the icon **e**

Exam-style questions

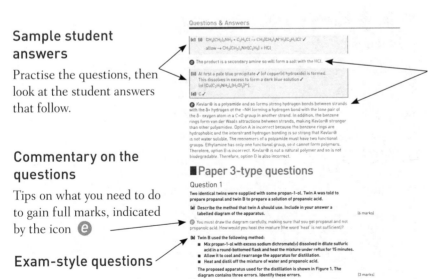

About this book

This guide is the fourth of a series covering the Edexcel specification for A-level chemistry. It offers advice for the effective study of topics 16–19. These four topics (together with topics 2 and 5, covered in the first student guide of this series, and topics 6, 7 and 9, covered in the second student guide) are examined on A-level paper and as part of synoptic paper 3 examination, which draws on all of the topics 1–19. The guide has two sections:

The **Content Guidance** is not intended to be a textbook. It offers guidelines on the main features of the content of topics 16–19, together with particular advice on practical chemistry that will be tested in paper 3.

The **Questions & Answers** section starts with an introduction that gives advice on approaches and techniques to ensure that you answer the exam questions in the best way you can. It then provides exam-style questions with student answers and comments on how the answers would be marked.

The specification

The specification states the chemistry that can be examined at A-level and describes the format of the tests. It can be obtained from Edexcel, either as a printed document or from the web at www.edexcel.com.

Learning to learn

Learning is not instinctive — you have to develop techniques to make good use of your time. Chemistry has specific difficulties that need to be understood if your studies are to be effective from the start.

Reading chemistry textbooks

Chemistry textbooks are a valuable resource, not only for finding out the information for your homework but also to help you understand concepts of which you are unsure. They need to be read carefully, with a pen and paper to hand for jotting down things as you go — for example, making notes, writing equations, doing calculations and drawing diagrams. Reading and revising are *active* processes that require concentration. Looking vaguely at the pages is a waste of time. In order to become fluent and confident in chemistry, you need to master detail.

Chemical equations

Equations are quantitative, concise and internationally understood. When you write an equation, check that:

- you have thought of the *type* of reaction occurring — for example, is it neutralisation, addition or disproportionation?
- you have written the correct formulae for all the substances
- your equation balances both for the numbers of atoms of each element and for charge
- you have not written something silly, such as having a strong acid as a product when one of the reactants is an alkali
- you have included *state symbols* in all thermochemical equations if they have been asked for

Diagrams

Diagrams of apparatus should be drawn in section. When you see them, copy them and ask yourself why the apparatus has the features it has. What is the difference between a distillation and a reflux apparatus, for example? When you do practical work, examine each piece of the apparatus closely so that you know both its form and function.

Calculations

Calculations are not normally structured at A-level. Therefore, you will need to *plan* the procedure for turning the data into an answer.

- Set your calculations out fully, making it clear what you are calculating at each step. Don't round figures up or down during a calculation. Either keep all the numbers on your calculator or write any intermediate answers to four significant figures.
- If you have time, check the accuracy of each step by recalculating it. It is easy to enter a wrong number into your calculator, or to calculate a molar mass incorrectly.
- Think about the size of your final answer. Is it far too big or vanishingly small?
- Finally, check that you have the correct *units* in your answer and that you have given it to an appropriate number of *significant figures* — if in doubt, give it to three or, for pH calculations, to two decimal places.

Notes

Most students keep notes of some sort. Notes can take many forms: they might be permanent or temporary; they might be lists, diagrams or flowcharts. You have to develop your own styles. For example, notes that are largely words can often be recast into charts or pictures, which is useful for imprinting the material. The more you rework the material, the clearer it will become.

Whatever form your notes take, they must be organised. Notes that are not indexed or filed properly are useless, as are notes written at enormous length and those written so cryptically that they are unintelligible a month later.

Writing

You need to be able to write concisely and accurately. This requires you to marshal your thoughts properly and needs to be practised during your ordinary learning.

There are no marks specifically for 'communication skills', but if you are not able to communicate your ideas clearly and accurately, then you will not score full marks. The space available for an answer is a poor guide to the amount that you have to write — handwriting sizes differ hugely, as does the ability to write crisply. Filling the space does not necessarily mean you have answered the question. The mark allocation suggests the number of points to be made, not the amount of writing needed.

Content Guidance

▌Topic 16 Kinetics II

Required year 1 knowledge

Collision theory

The following are important factors:

- Collision frequency — how often the molecules collide in a given time.
- Collision energy — particles must collide with enough kinetic energy to cause a reaction. The minimum energy that two molecules must have on collision in order to react is called the activation energy, E_a.
- Orientation on collision — no reaction will occur if the OH⁻ ion collides with the CH_3 group in the nucleophilic substitution reaction between, for example, chloroethane and hydroxide ions. The collision must occur with the carbon carrying the chlorine.

Maxwell–Boltzmann distribution

The molecules in a gas or solution have a wide range of kinetic energies. This is shown graphically in the Maxwell–Boltzmann distribution at two temperatures T_{cold} and T_{hot} (Figure 1).

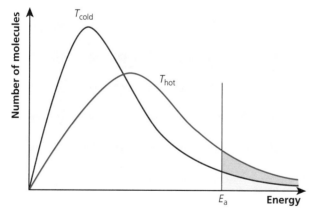

Figure 1 Maxwell–Boltzmann distribution at two temperatures

difference between the two curves is that the curve for T_{hot} has:
- lower modal value (the peak)
- modal value to the right of that of T_{cold}

Exam tip

You are expected to know the kinetics you studied in year 1 of the A-level course. In particular, you should have a thorough understanding of collision theory, including the Maxwell–Boltzmann distribution of molecular energies.

Exam tip

Note that both graphs start at the origin, but neither touches the x-axis at high energies.

Effect of temperature on rate

The *area* under the curve to the *right* of the activation energy (shaded on the graphs in Figure 1) is a measure of the number of molecules that have enough energy to react on collision. This area is greater on the T_{hot} graph compared with the T_{cold} graph. Therefore, the number of collisions between molecules with enough energy to react is greater. This means that a greater *proportion* of the collisions result in reaction, so the rate of reaction is faster at the higher temperature.

Effect of catalyst on rate

A catalyst causes the reaction to proceed by an *alternative route* that has a *lower* activation energy.

In Figure 2, the value of E_{cat} is less than that of E_a, so the area under the curve to the right of E_{cat} is greater than the area to the right of E_a. This means that a greater *proportion* of molecules have the required lower activation energy on collision, allowing reaction to occur.

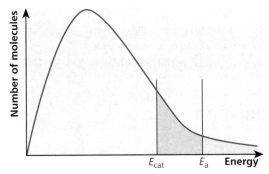

Figure 2 Graph showing the effect of a catalyst on the number of molecules with the required activation energy

Effect of concentration and pressure on rate

An increase in the concentration of a reactant in solution, or in the pressure for gaseous reactions, increases the *frequency* of collisions. Therefore, the reaction rate increases.

Year 2 kinetics

Definitions

■ The **rate of reaction** is the amount by which the concentration changes in a given time. Its units are $mol\,dm^{-3}\,s^{-1}$. There are two ways of determining the rate:

  1 The initial rate of reaction is the rate of reaction at the time when the reactants are mixed. Its value can be estimated by measuring the concentration of a reactant at the start and after a short period of time. If the concentration has not changed by more than 10%:

$$\text{initial rate} = \frac{\text{change in concentration}}{\text{time}}$$

Exam tip

A reaction mixture i
kinetically stable if t
activation energy is
high that the reactio
too slow to be obser
at room temperatur

2 The concentration of a reactant or product is measured at regular time intervals and a graph of concentration against time is plotted. The slope (gradient) of this line at a given point is the rate of reaction at that point. The slope is calculated by dividing the *change* in concentration by the time difference (see page 13).

Knowledge check 1

When excess 2-iodopropane was mixed with a small amount of aqueous sodium hydroxide, the initial concentrations were:
$[CH_3CHICH_3] = 0.0028 \, mol \, dm^{-3}$ and $[OH^-] = 0.00020 \, mol \, dm^{-3}$
All the sodium hydroxide had reacted after 26 s. Calculate the initial rate of reaction.

The **rate equation** connects the rate of a reaction with the concentration of the reactants. For a reaction:

$$xA + yB \rightarrow products$$

the rate equation is:

$$rate = k[A]^m[B]^n$$

m and n may or may not be the same as x and y, and one or other of m and n could be zero. Note that if the reaction is carried out in the presence of a homogeneous catalyst, the concentration of the catalyst will appear in the rate equation.

The **order of reaction** is the sum of the powers to which the *concentrations* of the reactants are raised in the rate equation. In the example above, the order is $(m + n)$. The partial order with respect to one reactant is the power to which its concentration is raised in the rate equation. In the example above, the partial order with respect to A is m.

The **rate constant**, k, is the constant of proportionality that connects the rate of the reaction with the concentration of the reactants, as shown in the above rate equation. Its value alters with temperature. A reaction with a *large* activation energy has a *low* value of k.

The **half-life** of a reaction is the time taken for the concentration of a reagent to halve. This means that half of the reagent has reacted. The half-life of a first-order reaction is constant. The half-life of a second-order reaction increases as the concentration falls.

In a multi-step reaction, the **rate-determining step** is the slowest step and controls the overall rate of the reaction. The reagents that appear in the rate equation are those that are involved in the rate-determining step and any previous steps. Any substance that only appears in the mechanism *after* the rate-determining step will not appear in the rate equation.

The **activation energy**, E_a or E_{act}, of a reaction is the minimum energy that colliding molecules must have in order for that collision to result in a reaction.

Catalysts can be described as either heterogeneous or homogeneous. A homogeneous catalyst is in the *same phase* as the reactants. An example is aqueous iron(II) ions in the oxidation of aqueous iodide ions by aqueous persulfate ions. A heterogeneous catalyst is in a *different* phase from the reactants. An example is the solid iron catalyst in the reaction between gaseous nitrogen and hydrogen to produce ammonia (the Haber process).

Knowledge check 2

The rate equation for a reaction between A and B was found to be:

$$rate = k[A][B]$$

What are the units of k?

Knowledge check 3

A reaction between A and B to form P is catalysed by C and takes place in two steps:

slow: $A(aq) + C(aq) \rightarrow X(aq)$

fast: $B(aq) + X(aq) \rightarrow P(aq) + C(aq)$

Write the rate equation for this reaction.

Experimental techniques

All the methods described below must be carried out at *constant temperature*, preferably using a thermostatically controlled water bath. The laboratory itself is often considered as a temperature-controlled environment, but this can cause considerable inaccuracy if the reaction is significantly exothermic.

Continuous monitoring

These methods can be used to gather data and then plot a graph of concentration against time, see pages 12–14.

Titration by sampling

- Mix the reactants, stir and start timing.
- Remove samples at regular intervals and quench the reaction by adding the samples to iced water or a substance that will react with the catalyst or one of the reactants.
- Titrate samples against a suitable solution.

This method can be used when a reactant or product is:

- an acid — titrate against an alkali
- a base — titrate against an acid
- iodine — titrate against sodium thiosulfate solution, using starch as an indicator

Core practical 13a

Iodination of propanone

In acid solution iodine reacts with propanone to form iodopropanone and hydrogen iodide:

$$CH_3COCH_3 + I_2 \xrightarrow{H^+(aq)} CH_3COCH_2I + HI$$

This reaction can be monitored by the following method.

- Place $25\,cm^3$ of iodine solution of known concentration in a flask and add $25\,cm^3$ of sulfuric acid solution (an excess) and $50\,cm^3$ of water. Add $5\,cm^3$ of pure propanone from a burette, starting the clock as you do so.
- At noted intervals of time remove $10\,cm^3$ of the reaction mixture and quench it by adding it to a slight excess of aqueous sodium hydrogencarbonate. (This reacts with the acid catalyst, stopping the reaction.)
- Quickly titrate the remaining iodine against standard sodium thiosulfate solution, using starch as an indicator.
- Repeat by removing samples at regular intervals.

The rate equation is of the form:

$$\text{rate} = k[\text{propanone}]^x[H^+]^y[I_2]^z$$

Because a large excess of propanone was taken, the concentration of propanone remains approximately constant. Even though acid is produced in the reaction, the original amount of acid catalyst was large and so $[H^+]$ remains effectively constant. This means that the rate equation becomes:

$$\text{rate} = k'[I_2]^z \text{ where } k' = k[\text{propanone}]^x[H_+]^y$$

→

Knowledge check

Why must sodium hydroxide *not* be add to quench the reactic

The volume of sodium thiosulfate is proportional to the amount of iodine and hence to the iodine concentration. So a graph of volume of sodium thiosulfate against time will have the same shape as the graph of $[I_2]$ against time.

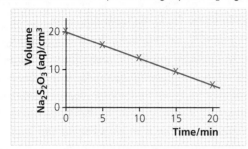

Figure 3 Graph showing volume of sodium thiosulfate over time

As this graph is a straight line, the slope and hence the rate are constant, so the reaction is zero order with respect to iodine. A conclusion that can be drawn from this is that iodine must enter the mechanism *after* the rate-determining step.

If the experiment is repeated with twice as much propanone, the slope of the linear graph of $[I_2]$ against time will be twice that of the slope of the former graph. This shows that k' has doubled, so the reaction must be first order with respect to propanone.

Similarly, if a third set of experiments is carried out varying the concentration of acid catalyst, the rate will double as the acid concentration doubles. This shows that the reaction is also first order with respect to the acid catalyst.

Exam tip

The mechanism must first involve acid and propanone in a rate-determining slow step, producing an intermediate. This is then followed by a faster step involving the intermediate reacting with iodine.

Colorimetry

If a reactant or product is coloured, follow the reaction in a colorimeter. The intensity of the colour is a measure of the concentration. The colorimeter has to be calibrated so that the transmittance can be converted to concentration.

Mass change or volume of gas

If a gas is produced, *either* measure the volume of gas at regular intervals (collecting it over water in a measuring cylinder or using a gas syringe) *or* carry out the reaction on a top pan balance and measure the loss in mass at regular intervals.

Exam tip

Be aware that carrying out the reaction on a top pan balance can be inaccurate as the mass of gas released will be very small.

Rotation of plane of polarisation

If a reactant is a single optical isomer, follow the reaction using a polarimeter. The angle of rotation depends on the concentration.

Initial rates

Fixed amount of product

Some reactions can be followed by measuring the time taken to produce a fixed amount of detectable product. The experiment is then repeated, changing the concentration of one of the reactants. In this type of experiment, the rate is calculated from the *reciprocal* of the time, i.e. 1/time is a measure of the rate. Some examples are:

Exam tip

The use of a pH meter to follow the change in acidity of a reaction is inaccurate.

- the reaction between dilute acid and aqueous sodium thiosulfate. The reactants are mixed in a beaker standing on a tile marked with a large X. The time taken for enough sulfur to be precipitated to hide the X is measured.
- the reaction between dilute acid and a metal such as magnesium or a carbonate such as calcium carbonate. The time taken for a fixed volume of gas is measured and the rate is proportional to 1/time.

If the acid is in considerable excess, the time for all the magnesium to disappear is measured and 1/time is approximately proportional to the rate.

Clock methods

In a 'clock' reaction, the reactants are mixed and the time taken to produce a fixed amount of product is measured. The experiment is then repeated several times using different starting concentrations. This gives several initial rates of reaction at different concentrations.

Alternatively, the experiment can be carried out with one set of concentrations, but at different temperatures. This enables the activation energy to be calculated (see page 20).

Core practical 13b

The iodine 'clock'
The oxidation of iodide ions by hydrogen peroxide in acid solution can be followed as a 'clock' reaction:

$$H_2O_2(aq) + 2I^-(aq) + 2H^+(aq) \rightarrow I_2(s) + 2H_2O(l)$$

- 25 cm^3 of hydrogen peroxide solution is mixed in a beaker with 25 cm^3 of water and a few drops of starch solution are added.
- 25 cm^3 of potassium iodide solution and 5 cm^3 of a dilute solution of sodium thiosulfate are placed in a second beaker.
- The contents of the two beakers are mixed and the time taken for the solution to go blue is measured.
- The experiment is repeated with the same volumes of potassium iodide and sodium thiosulfate but with 20 cm^3 of hydrogen peroxide and 30 cm^3 of water, and then with other amounts of hydrogen peroxide and water, totalling 50 cm^3.

The hydrogen peroxide reacts with the iodide ions to produce iodine, which immediately reacts with the thiosulfate ions. When all the thiosulfate ions have been used up, the next iodine will produce an intense blue-black colour with the starch. At this point the clock is stopped. The rate of reaction is approximately proportional to 1/time. As the concentration of the reactants hardly alters during the course of a measurement, 1/time is also proportional to the initial rate.

Interpretation of kinetic data

Concentration–time or volume–time graphs

The data obtained by repeated sampling involving titration of a reactant or product can be used to plot a graph of titre volume against time (see Core practical 13a on page 10). Continuous monitoring methods, such as colorimetry, give data about concentration at intervals of time.

Exam tip

This method is only accurate if 10% or less of the reactant has reacted when the fixed amount of product is observed.

Exam tip

The clock can be 'rewound' by adding a second amount of sodium thiosulfate and adding the second time taken to the first.

The most usual graphs have the titre volume or the concentration of a reactant on the *y*-axis and time on the *x*-axis. The slope of this type of graph at any point is the rate of the reaction at that point.

Example 1

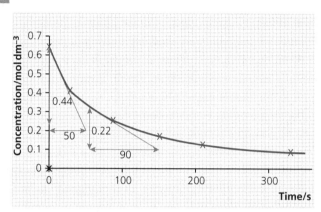

Figure 4 Graph showing concentration over time for an unspecified reaction

In Figure 4, the slope at the initial concentration of $0.64 \, \text{mol dm}^{-3}$ = $(0.64 - 0.20)/50$ = $0.0088 \, \text{mol dm}^{-3} \, \text{s}^{-1}$.

The slope at $0.32 \, \text{mol dm}^{-3}$ = $(0.32 - 0.10)/(150 - 60)$ = $0.0024 \, \text{mol dm}^{-3} \, \text{s}^{-1}$.

This value is close to ¼ that of the initial rate. As the rate decreased by a factor of 4 when the concentration halved, the reaction is second order.

Example 2

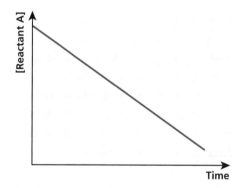

Figure 5 Graph showing concentration over time for an unspecified reaction

In Figure 5, the slope is constant, therefore the rate is constant, and the reaction is zero order with respect to A.

Exam tip

This is the type of graph obtained in the iodination of propanone using a large excess of *both* propanone and the acid catalyst. The reaction is, therefore, zero order with respect to iodine.

Example 3

The half-life is the time taken for the *concentration of a reactant to halve*. For a first-order reaction, the half-life is *constant*. Its value can be determined from a graph of concentration against time.

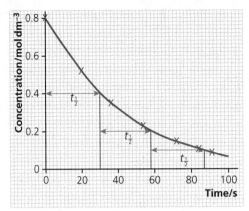

Figure 6 Graph showing concentration over time for an unspecified reaction

In Figure 6, the time taken for the concentration to halve from 0.8 to $0.4 \, mol \, dm^{-3}$ is 30 s.

The time taken for the concentration to halve from 0.4 to $0.2 \, mol \, dm^{-3}$ is 28 s.

The time taken for the concentration to halve from 0.2 to $0.1 \, mol \, dm^{-3}$ is 29 s.

The half-lives are *constant* to within experimental error and so the reaction is first order. The average half-life is $(30 + 28 + 29)/3 = 29 \, s$.

Exam tip

The second half-life is the time for the concentration to fall from 0.4 to $0.2 \, mol \, dm$ (28 s) and *not* the time for the concentration to fall from 0.8 to 0.2 (58 s). This is a common error.

Example 4

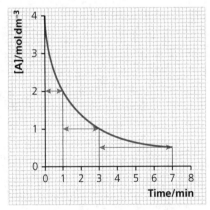

Figure 7 Graph showing concentration over time for an unspecified reaction

The first half-life is 1 minute. The second successive half-life is 2 minutes and the third is 4 minutes. As the half-lives are doubling as the concentration halves, the reaction is second order with respect to A.

Exam tip

If the half-life is constant, the reactio is first order. If it increases, the reaction is second or a higher order.

Rate–concentration graphs

The most usual graphs have the rate of reaction on the *y*-axis and concentration of a reactant or the concentration of a reactant squared on the *x*-axis. If a straight line is obtained, the rate of reaction is proportional to the quantity plotted on the *x*-axis.

For gases, the partial pressure rather than the concentration can be plotted. The partial pressure of a gas is proportional to its concentration.

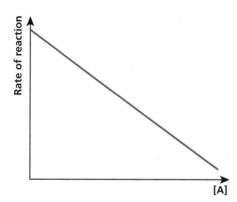

Figure 8 Graph showing rate of reaction versus concentration of reactant for an unspecified reaction

In Figure 8, the graph is a straight line, so the rate is proportional to the concentration of A. This means that the reaction is first order with respect to A.

Example 2

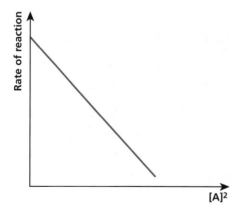

Figure 9 Graph showing rate of reaction versus concentration of reactant squared, for an unspecified reaction

In Figure 9, this graph is a straight line, so the rate is proportional to the *square* of the concentration of A. This means that the reaction is second order with respect to A.

Deducing the order of reaction from initial rate data

To do this, you need data from at least three experiments. For example, data from the reaction below could be used:

$$xA + yB \rightarrow products$$

Table 1 Rate data from three experiments

Experiment	[A] /mol dm^{-3}	[B] /mol dm^{-3}	Initial rate /mol dm^{-3} s^{-1}
1	0.1	0.1	p
2	0.2	0.1	q
3	0.2	0.2	r

Comparing experiments 1 and 2 shows that [A] has doubled but [B] has stayed the same.

If the rate is unaltered ($q = p$), the order with respect to A is 0.

If the rate doubles ($q = 2p$), the order with respect to A is 1.

If the rate quadruples ($q = 4p$), the order with respect to A is 2.

The order with respect to B can be determined in a similar way, by analysing experiments 2 and 3.

Exam tip

When finding the order of each reactant you should always state which pair of experiments you are using.

Worked example 1

In Table 1, suppose $p = 0.0024$, $q = 0.0096$ and $r = 0.0096$.
1. Determine the partial orders of A and B.
2. Determine the total order.
3. Write the rate equation.
4. Calculate the value of the rate constant.

Answer

1. From experiments 1 and 2: when [A] doubles, the rate quadruples ($0.0096 = 4 \times 0.0024$). Therefore, the order with respect to A is 2.
2. From experiments 2 and 3: when [B] doubles, the rate is unaltered. Therefore, the order with respect to B is zero. The total order is $2 + 0 = 2$.
3. The rate equation is:
   $$rate = k[A]^2$$
4. The rate constant is:
   $$k = \frac{rate}{[A]^2} = \frac{0.0024\,mol\,dm^{-3}\,s^{-1}}{(0.1\,mol\,dm^{-3})^2} = 0.24\,mol^{-1}\,dm^3\,s^{-1}$$

Calculation of the rate constant

You must first find the order of reaction.

From rate data

If you know the rate of reaction between substances A and B at given concentrations and the order of reaction with respect to A and B, the rate constant can be found using the equation:

k = rate of reaction/$[A]^m[B]^n$

where m and n are the known orders.

Rate constant units

Table 2 Rate constant units for each order

Total order	Units of k
Zero	$mol\,dm^{-3}\,s^{-1}$
First	s^{-1}
Second	$mol^{-1}\,dm^3\,s^{-1}$

Order and mechanisms

The experimentally determined partial orders of a reaction provide evidence for the mechanism of the reaction.

The iodination of propanone

Core practical 13a (see page 10) showed that this reaction is zero order with respect to iodine.

Repeating the experiment with different initial amounts of propanone and acid shows that it is first order in propanone and also in the acid catalyst. The following conclusions can be drawn:

The rate equation is:

$$rate = k[propanone]^1[H^+]^1[I_2]^0$$

- Iodine enters the mechanism *after* the rate-determining step.
- Both propanone and H^+ appear once in the mechanism before or in the rate-determining step.

Based on this information the following mechanism can be suggested as being compatible with the experimental data.

Step 1: The lone pair of electrons on the oxygen forms a bond with an H^+ from the catalyst, as shown in Figure 10.

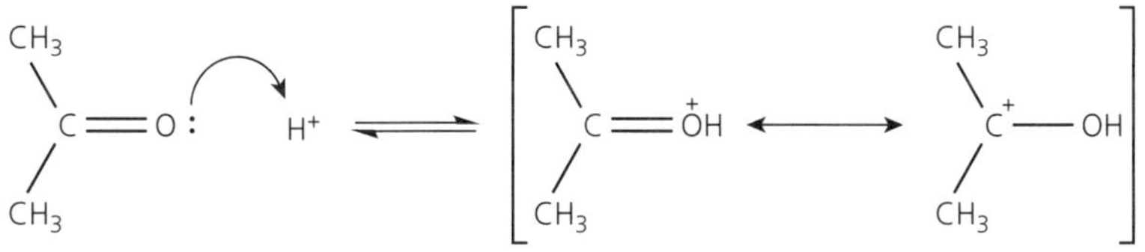

Figure 10 A lone pair of electrons on the oxygen forms a bond with an H^+

Step 2: Formation of the enol intermediate, as shown in Figure 11. This is slow because a C–H bond has to be broken and hence is the rate-determining step.

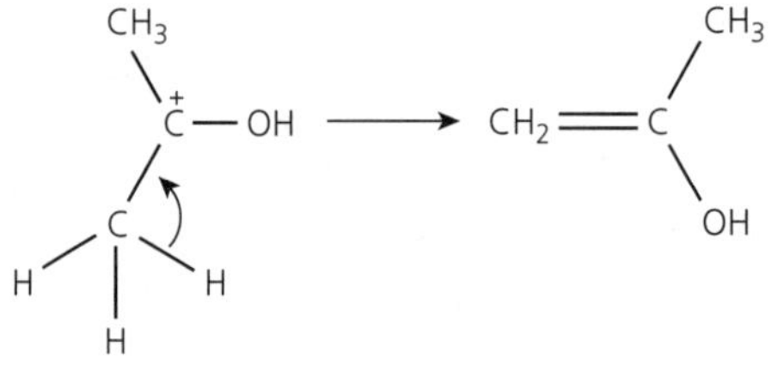

Figure 11 Formation of the enol intermediate

Step 3: Addition of iodine and loss of H^+ as shown in Figure 12. This is a fast step.

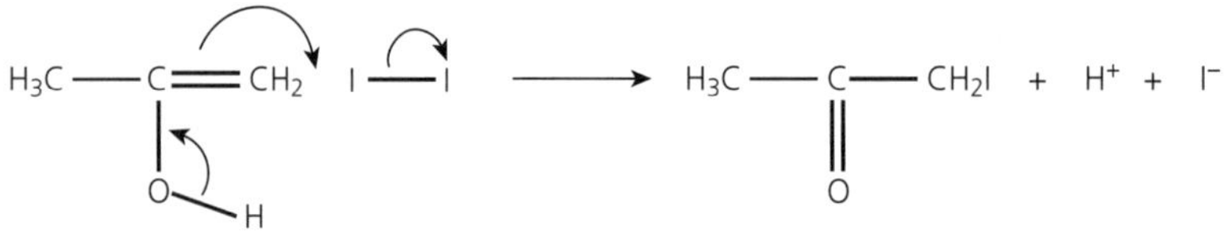

Figure 12 Addition of iodine and loss of H^+

The hydrolysis of halogenoalkanes

The overall reaction is:

$$R–X + OH^- \rightarrow ROH + X^-$$

where X is a halogen and R an alkyl group such as C_2H_5 or $(CH_3)_2CH$.
With primary halogenoalkanes the rate equation is:

$$\text{rate} = k[RX]^1[OH^-]^1$$

With a tertiary halogenoalkane it is:

$$\text{rate} = k[RX]^1[OH^-]^0$$

Nucleophilic substitution second order

The S_N2 mechanism involves both the halogenoalkane and OH^- in the rate-determining step, as shown in Figure 13.

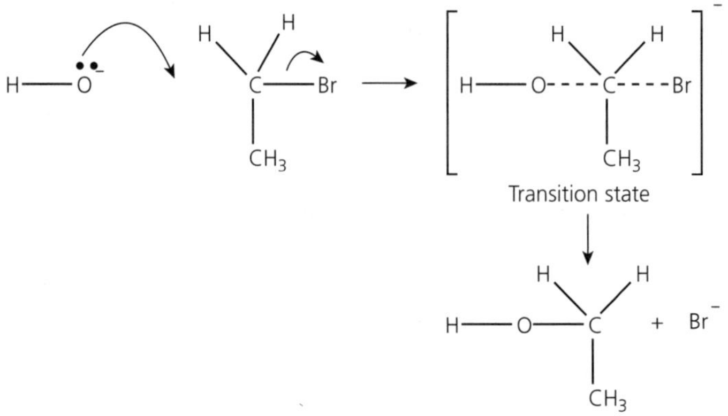

Transition state

Figure 13 The rate-determining step involving the halogenoalkane and nucleophile, proceeding via a transition step

Exam tip

In the transition state there are five groups around the central carbon atom. This would cause considerable steric strain in secondary halogenoalkanes and even more so in tertiary halogenoalkanes.

Nucleophilic substitution first order

S_N1 is a two-step reaction with the hydroxide ion only appearing in the second and faster step, as shown in Figure 14.

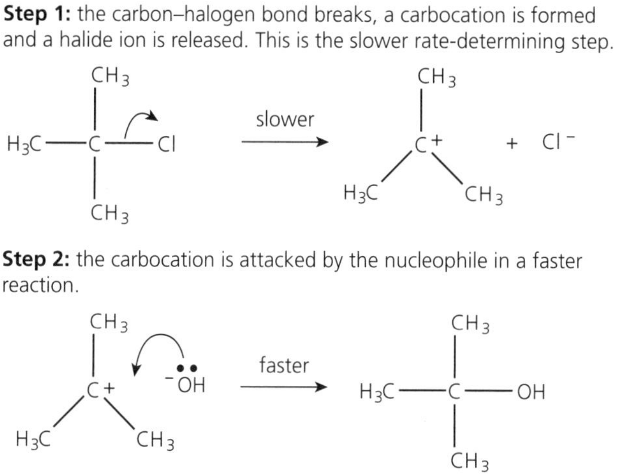

Step 1: the carbon–halogen bond breaks, a carbocation is formed and a halide ion is released. This is the slower rate-determining step.

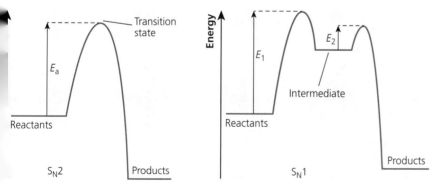

Step 2: the carbocation is attacked by the nucleophile in a faster reaction.

Figure 14 The rate-determining step involving the halogenoalkane and not the nucleophile, proceeding via an intermediate

Primary halogenoalkanes react almost totally by an S_N2 mechanism, but tertiary halogenoalkanes react almost entirely by the S_N1 route. The rate of S_N2 decreases from primary to tertiary as the nucleophile is increasingly blocked by alkyl groups (steric effect). The rate of S_N1 increases from primary to tertiary because the intermediate planar carbocation is stabilised by 'electron-pushing' groups, such as the CH_3 group.

The reaction-profile diagrams for these two different types of reaction are shown in Figure 15.

Figure 15 Reaction profile diagrams for S_N2 and S_N1

Note that as E_1 is greater than E_2, step 1 is rate-determining.

Effect of temperature on rate

An increase in temperature *always* increases the rate of a reaction because the rate constant increases with temperature. The mathematical relationship is given by the Arrhenius equation:

Exam tip

The intermediate is planar with three groups around the positive carbon atom. Therefore, there would be no steric strain even with three alkyl groups.

Exam tip

You must make sure that a curly arrow starts on a lone pair of electrons or on a bond between two atoms.

$k = Ae^{-E_a/RT}$ or $\ln k = \ln A - E_a/RT$

where A is a constant, E_a is the activation energy, R is the gas constant and T is the temperature in kelvin, where $25°C = (25 + 273)\,K$.

An increase in temperature causes E_a/RT to become smaller. Thus $-E_a/RT$ becomes less negative and hence $\ln k$ and the rate increase.

A change in temperature does *not* alter the value of the activation energy. However, the Arrhenius equation also shows the relationship between activation energy and rate constant. A larger value of E_a gives a more negative exponential power and hence a smaller value of k. A *higher* activation energy results in a *slower* reaction.

A catalyst provides an alternative route with a lower activation energy. The *lower* E_{cat} results in a *larger* k and hence a *faster* reaction.

Exam tip

You do not need to learn this equation because you will always be given it in th question.

Knowledge check 5

The C–Cl bond enthal is larger than the C–I bond enthalpy. Use th information to explain why iodoalkanes are hydrolysed at a faster rate than similar chloroalkanes.

Core practical 14

Calculating the activation energy
The correct way to do this is to measure the value of the rate constant, k, at different temperatures. A graph is then plotted of $\ln k$ against 1/temperature as shown in Figure 16, where the temperature must be in kelvin. The slope of this graph $= -E_a/R$, so $E_a = -(\text{slope} \times R)$. The gas constant, R, has a value of $8.31\,J\,K^{-1}\,mol^{-1}$.

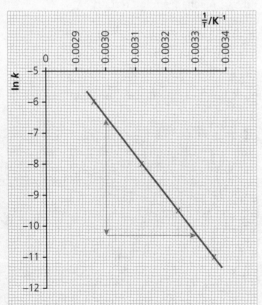

Figure 16 A graph of $\ln k$ against 1/temperature

$$\text{slope (gradient)} = \frac{-10.3 - (-6.5)}{(0.00330 - 0.00300)} = \frac{-3.8}{0.00030} = -1.27 \times 10^4\,K$$

$$E_a = -\text{slope} \times R = -(-1.27 \times 10^4 \times 8.31) = +1.05 \times 10^5\,J\,mol^{-1} = +105\,kJ\,mol^{-1}$$

Be careful about signs. The slope is calculated as $(y_2 - y_1)/(x_2 - x_1)$, where y_2 is the second value on the y-axis and x_2 the corresponding value on the x-axis.

Exam tip

Beware — the slope the graph is *negative* so $-(\text{slope} \times R)$ will b positive.

Summary

After studying this topic, you should be able to:

- understand the terms rate constant, half-life and activation energy
- know the effect of a change of temperature on the rate constant
- draw Maxwell–Boltzmann distribution curves and use them to explain the effect of a change of temperature and of adding a catalyst
- describe methods for following a reaction involving iodine, the production of a gas or where one substance is coloured

- deduce the order of reaction from initial rate data, from rate–concentration graphs and from consecutive half-lives
- explain why the iodination of propanone is zero order only when there is a large excess of propanone and acid
- calculate the activation energy from a graph of $\ln k$ against $1/t$
- relate the mechanism to the order of reaction

Topic 17 Organic chemistry II

Required year 1 knowledge

Nomenclature

You should refresh your knowledge of how organic compounds are named, using Table 3.

Table 3 Stem name used for organic compounds with different carbon chain lengths

Carbon chain length	Stem name
One atom	Meth-
Two atoms	Eth-
Three atoms	Prop-
Four atoms	But-
Five atoms	Pent-

If the chain contains three or more carbon atoms, the position of a functional group is indicated by a number — for example, $CH_3CH_2CH=CH_2$ is but-1-ene; $CH_3CH(OH)CH_3$ is propan-2-ol.

If the carbon chain is *branched*, the position of the alkyl branch is indicated by a number. For example, $CH_3CH_2CH(CH_3)CH_2CH_2OH$ is 3-methylpentan-1-ol.

Bond enthalpy and polarity

The smaller the value of the bond enthalpy, the weaker the bond. This means that the activation enthalpy of a reaction involving the breaking of that bond will be lower and the rate of reaction faster. For example, the C–I bond is weaker than the C–Cl bond, so iodoalkanes react faster than chloroalkanes.

The π-bond in alkenes is weaker than the σ-bond in alkanes, so alkenes are more reactive than alkanes.

The bond polarity determines the type of reaction. The carbon atom in halogenoalkanes is δ+, so it is attacked by nucleophiles.

AS reactions, reagents and conditions

Alkanes

In the presence of light, alkanes react with chlorine — for example:

$$CH_4 + Cl_2 \rightarrow CH_3Cl + HCl$$

The organic product is chloromethane.

Alkenes

In Table 4, propene is used as the example.

Table 4 Alkene reaction details

Reagent	Equation	Conditions	Organic product
Hydrogen	$CH_3CH=CH_2 + H_2 \rightarrow CH_3CH_2CH_3$	Heated nickel catalyst	Propane
Halogens e.g. Br_2	$CH_3CH=CH_2 + Br_2 \rightarrow CH_3CHBrCH_2Br$	Organic solvent	1,2-dibromopropane
Hydrogen halides e.g. HCl	$CH_3CH=CH_2 + HCl \rightarrow CH_3CHClCH_3$	Mix gases at room temperature	2-chloropropane
Oxidation e.g. $KMnO_4$	$CH_3CH=CH_2 + [O] + H_2O \rightarrow CH_3CH(OH)CH_2OH$	Aqueous solution	Propane-1,2-diol

Knowledge check 6

Define the terms *electrophile* and *nucleophile*.

Halogenoalkanes

In Table 5, 1-bromoethane is used as the example.

Table 5 Halogenoalkane reaction details

Reagent	Equation	Conditions	Organic product
Aqueous alkali e.g. NaOH(aq)	$CH_3CH_2Br + NaOH \rightarrow CH_3CH_2OH + NaBr$	Heat under reflux	Ethanol
Ethanolic alkali e.g. KOH	$CH_3CH_2Br + KOH \rightarrow CH_2=CH_2 + KBr + H_2O$	Heat under reflux in a solution of ethanol	Ethene
Water in the presence of silver nitrate solution	$CH_3CH_2Br + H_2O \rightarrow CH_3CH_2OH + H^+ + Br^-$ Then $Ag^+ + Br^- \rightarrow AgBr$	Add to silver nitrate solution and observe ppt of AgBr	Ethanol
Ammonia	$CH_3CH_2Br + 2NH_3 \rightarrow CH_3CH_2NH_2 + NH_4Br$	Use ammonia dissolved in ethanol	Ethylamine

Exam tip

You must know the mechanisms for S_N1 and S_N2 reactions of halogenoalkanes.

lcohols

Table 6, ethanol is used as the example.

ble 6 Alcohol reaction details

eagent	Equation	Conditions	Organic product
hosphorus entachloride	$CH_3CH_2OH + PCl_5 \rightarrow CH_3CH_2Cl + POCl_3 + HCl$	Dry at room temperature	Chloroethane
ydrogen bromide	$CH_3CH_2OH + HBr \rightarrow CH_3CH_2Br + H_2O$	HBr is made by adding 50% sulfuric acid to solid potassium bromide	Bromoethane
xidation e.g. with cidified potassium chromate(VII)	$CH_3CH_2OH + [O] \rightarrow CH_3CHO + H_2O$	Heat and distil out the aldehyde	Ethanal
	$CH_3CH_2OH + 2[O] \rightarrow CH_3COOH + H_2O$	Heat under reflux	Ethanoic acid

Exam tip

You must know the conditions and products of the reactions of alcohols with acidified potassium dichromate(VI).

omerism

mers have the same molecular formula, but the atoms are arranged differently hin the molecule.

ructural isomerism

Carbon chain — the isomers have different carbon chain lengths. For instance, butane ($CH_3CH_2CH_2CH_3$) and methylpropane ($CH_3CH(CH_3)CH_3$) have the same molecular formula, C_4H_{10}.

Positional — the same functional group is in a different position in the isomers, for example propan-1-ol ($CH_3CH_2CH_2OH$) and propan-2-ol ($CH_3CH(OH)CH_3$).

Functional group — the isomers are members of different homologous series. Examples are:

- propanoic acid (CH_3CH_2COOH) and methyl ethanoate (CH_3COOCH_3)
- ethanol (C_2H_5OH) and methoxymethane (CH_3OCH_3)

ometric isomerism

ometric (*cis–trans*) isomerism is a form of **stereoisomerism**. In an alkene, metric isomerism is the result of restricted rotation about a carbon–carbon double d, provided that the two groups on each atom of the C=C group are different from n other. This is shown in Figure 17.

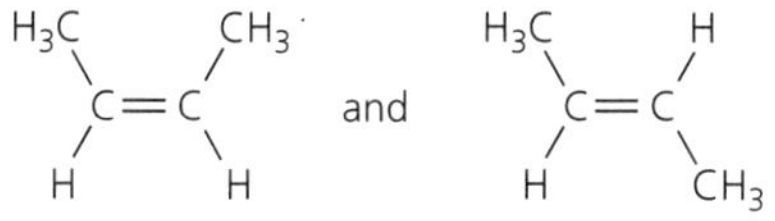

Figure 17 Geometric isomerism

Exam tip

If you are asked to draw a *displayed* formula, you must draw each individual atom and bond.

Knowledge check 7

Identify by name or formulae all the alkene isomers of C_4H_8.

The π-overlap in a double bond is above and below the plane of the molecule. Therefore, it is not possible to rotate about the double bond without breaking the π-bond, which only happens at high temperatures. Hence, the *cis* and *trans* isomers are different.

If there are three or four different groups bonded to the carbon atoms in the C=C group, the *E/Z* system must be used.

- Work out the priorities of the four groups (the higher the atomic numbers of the elements in the group, the higher the priority).
- If the higher priority groups are on opposite sides, it is an *E*-isomer.
- If the higher priority groups are on the same side, it is a Z-isomer.

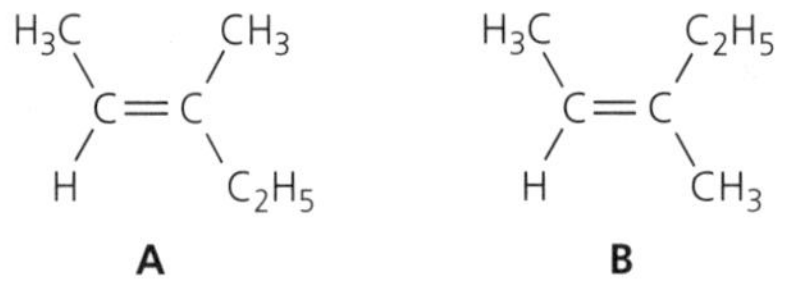

Exam tip

Z meanz on ze zame zide.

Figure 18 An *E*-isomer (compound A) and a *Z*-isomer (compound B)

In Figure 18, the higher priority on the left-hand carbon is the CH_3 group and the higher on the right-hand carbon is the C_2H_5 group.

Thus compound A in Figure 18 is (*E*)-3-methylpent-2-ene, as the two higher priority groups are on opposite sides, and compound B is (Z)-3-methylpent-2-ene, as the two higher priority groups are on the same side.

Knowledge check

Draw the *E*-isomer of 1-chloro-3-methylpe 2-ene.

Topic 17A Chirality

Optical isomerism

Optical isomerism is another form of stereoisomerism. It is the result of four different groups being attached to a carbon atom. This carbon atom is called the **chiral centre** and results in **chirality**. A chiral molecule is defined as a molecule that is non-superimposable on its mirror image.

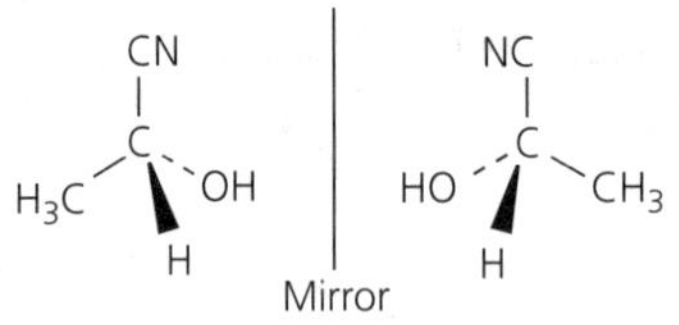

Figure 19 Optical isomers

Optical isomers can be distinguished from each other because they rotate the plane of plane-polarised light in opposite directions. A **racemic** mixture consists of equal amounts of two optical isomers which are mirror images of each other. Such a mixture does not have any effect on polarised light. Some chemical reactions result in a **racemic** (50:50) mixture of the two optical isomers.

Exam tip

You must draw the optical isomers as mirror images of each other, with wedges and with dot or dashes to give a three-dimensional appearance to your drawing.

Stereospecificity of nucleophilic substitution of halogenoalkanes

The mechanism for these reactions can be S_N1 or S_N2 (see pages 19 and 18). If a single optical isomer reacts with a nucleophile, such as hydroxide ions, the type of mechanism can be identified by the effect the final solution has on plane-polarised light.

The S_N1 mechanism proceeds through a planar intermediate. This can be attacked either from the top or the bottom and so a racemic mixture containing equal amounts of the two optical isomers will be produced. Thus the product will have no effect on the plane of polarisation of the plane-polarised light.

However, in the S_N2 mechanism, the nucleophile attacks from the side opposite to the halogen. This will result in the production of a single optical isomer (with inverted configuration), and so the solution will rotate the plane of polarisation of the polarised light.

The effect, if any, of the product on plane polarised light is compelling evidence for the type of mechanism.

Topic 17B Carbonyl compounds

Carbonyl compounds contain the C=O group. There are two types of carbonyl compound.

Aldehydes have a hydrogen atom on the carbonyl carbon and so have a –CHO group. Examples include methanal (HCHO), ethanal (CH_3CHO) and propanal (CH_3CH_2CHO).

Ketones have *two* alkyl groups attached to the carbonyl carbon atom. Examples include propanone (CH_3COCH_3) and butanone ($CH_3COCH_2CH_3$).

Physical properties

Neither aldehydes nor ketones have a hydrogen atom that is sufficiently $\delta+$, so they cannot form intermolecular hydrogen bonds. This means that they have a lower boiling point than alcohols with a similar number of electrons.

However, they contain a $\delta-$ oxygen atom and its lone pair of electrons can form a hydrogen bond with the $\delta+$ hydrogen atoms in water. Thus lower members of both aldehyde and ketone homologous series are water soluble. The solubility decreases as the number of carbon atoms increases, due to the hydrophobic hydrocarbon tail.

Preparation

Aldehydes

An aldehyde is prepared by oxidising a *primary* alcohol and distilling off the aldehyde that is produced. For example, a mixture of potassium dichromate(VI) and dilute sulfuric acid is added to hot ethanol and the ethanal distils off:

$$CH_3CH_2OH + [O] \rightarrow CH_3CHO + H_2O$$

Exam tip

The C=O group is polar, as oxygen is more electronegative than carbon. This means that the $\delta+$ carbon will be attacked by nucleophiles.

Exam tip

In both the reaction with aldehydes and the reaction with ketones (page 26), the colour changes from orange ($Cr_2O_7^{2-}$) to green (Cr^{3+}). Tertiary alcohols are not oxidised, so the solution would stay orange.

Ketones

Ketones are prepared by heating a *secondary* alcohol under reflux with an oxidising agent, such as a solution of potassium dichromate(VI) and dilute sulfuric acid:

$$CH_3CH(OH)CH_3 + [O] \rightarrow CH_3COCH_3 + H_2O$$

Reactions limited to aldehydes

Aldehydes contain the –CHO group, which can be oxidised to –COOH or to the –COO⁻ ion. Ketones do not have an easily removed hydrogen atom and so are *not* oxidised by the reagents below.

Reaction with Fehling's or Benedict's solution

When an aldehyde is warmed with Fehling's solution (or Benedict's solution), a red precipitate of copper(I) oxide is formed from the blue solution. The aldehyde is oxidised to a carboxylate ion:

$$CH_3CHO + [O] + OH^- \rightarrow CH_3COO^- + H_2O$$

Ketones have no effect on Fehling's or Benedict's, which stay blue. This is a test that can be used to distinguish between aldehydes and ketones.

Reaction with Tollens' reagent

When an aldehyde is added to Tollens' reagent and warmed, a silver mirror is formed. Ethanal is oxidised to the ethanoate ion:

$$CH_3CHO + [O] + OH^- \rightarrow CH_3COO^- + H_2O$$

Ketones do not form a silver mirror, and so this test can also be used to distinguish between an aldehyde and a ketone.

Reaction with acidified dichromate ions

When an aldehyde is warmed with a solution of potassium dichromate(VI) in sulfuric acid, the orange colour turns green and the aldehyde is oxidised to a carboxylic acid:

$$CH_3CHO + [O] \rightarrow CH_3COOH$$

Reactions common to aldehydes and ketones

Reduction

Aldehydes and ketones can be reduced by lithium aluminium hydride ($LiAlH_4$) in dry ether, *followed* by hydrolysis of the intermediate by dilute acid:

$$CH_3COCH_3 + 2[H] \rightarrow CH_3CH(OH)CH_3$$

This is an example of a nucleophilic addition reaction.

Reactions with nucleophiles

The carbon atom is δ+. Therefore, carbonyl compounds react with nucleophiles.

Exam tip

Fehling's and Benedict's solutions are complexes of copper(II) ions in *alkaline* solution. This means that the carboxylate anion and not the acid is produced.

Exam tip

Tollens' reagent is a complex between ammonia and silver(I) ions.

Vith hydrogen cyanide in the presence of a catalyst f cyanide ions

lthough the reactant is hydrogen cyanide, potassium cyanide in a buffer solution at H 8 is added to the carbonyl compound. This is necessary because the attack is by e nucleophile CN^-, followed by the addition of H^+ from HCN, so both CN^- ions d HCN need to be present.

r the reaction with ethanal, the organic product is 2-hydroxypropanenitrile:

$$CH_3CHO + HCN \rightarrow CH_3CH(OH)CN$$

r the reaction with propanone, the organic product is 2-hydroxy-2-ethylpropanenitrile:

$$CH_3COCH_3 + HCN \rightarrow CH_3C(CN)(OH)CH_3$$

e overall reactions are examples of nucleophilic addition.

e mechanism of this reaction, using ethanal as an example, is shown in Figure 20.

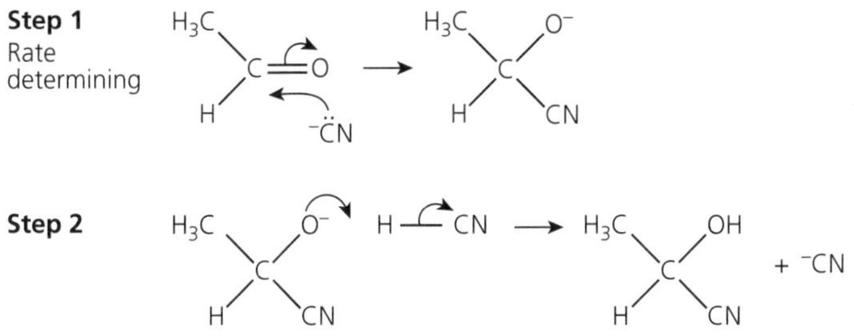

Figure 20 Mechanism of the reaction of hydrogen cyanide with ethanal

ereospecificity of nucleophilic addition of HCN

cause the original carbonyl compound is planar at the reaction site, the cyanide ion l attack equally from either above or below. This means that a racemic mixture of two enantiomers is formed when aldehydes or asymmetric ketones react (as shown Figures 21 and 22).

$CH_3CHO + HCN \rightarrow$ racemic mixture of

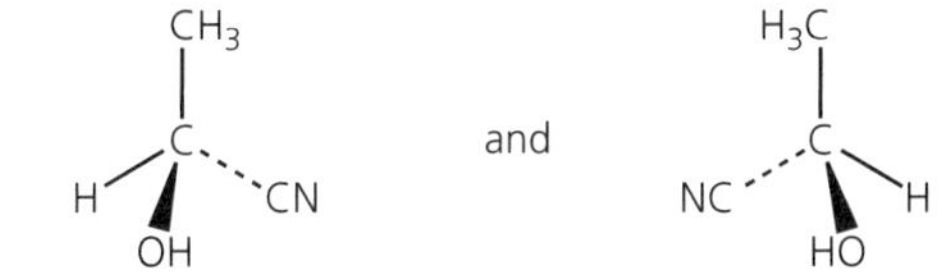

Figure 21 Two enantiomers formed in the reaction between an aldehyde and hydrogen cyanide

$CH_3COC_2H_5 + HCN \rightarrow$ racemic mixture of

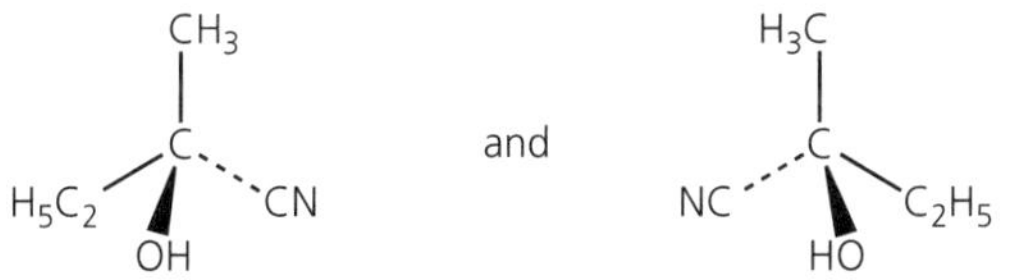

ure 22 Two enantiomers formed in the reaction between an asymmetric ketone and hydrogen cyanide

Exam tip

If the pH is too low, there are too few CN^- ions and so the reaction does not take place. If the pH is too high, there are not enough HCN molecules (or H^+ ions) and the reaction does not take place.

Exam tip

The curly arrow *must* start from the lone pair of electrons on the carbon atom of the CN^- group. Note that it is acceptable to have the curly arrow in step 2 going towards an H^+ ion rather than towards the H of an HCN molecule.

Exam tip

Do not confuse this reason with that involving S_N1 mechanisms. The addition to carbonyl compounds can result in a racemic mixture because the *starting* substance is planar, whereas the racemic mixture can be formed in S_N1 because the *intermediate* is planar.

Reaction with 2,4-dinitrophenylhydrazine

Both aldehydes and ketones form an orange-yellow precipitate with a solution of 2,4-dinitrophenylhydrazine (2,4-dnp):

$$(CH_3)_2\,C=O + H_2N-NH-C_6H_4(NO_2)_2 \rightarrow (CH_3)_2C=N-NH-C_6H_4(NO_2)_2 + H_2O$$

This is an example of a **nucleophilic substitution** reaction. It is used as a test — the formation of a precipitate confirms the presence of an aldehyde or a ketone.

This precipitate can be purified by recrystallisation (see page 54) and dried. Its melting temperature can then be measured and used to identify the aldehyde or ketone (Table 7).

Table 7 Melting temperatures of carbonyl compounds

Carbonyl compound	Melting temperature of 2,4-dnp derivative/°C	Carbonyl compound	Melting temperature of 2,4-dnp derivative/°C
Ethanal	168	Propanone	126
Propanal	150	Butanone	117
Butanal	123	Pentan-2-one	144
		Pentan-3-one	156

Iodoform reaction

Compounds with a $-COCH_3$ group undergo the following reaction when warmed gently with iodine and a few drops of alkali:

$$CH_3CHO + 3I_2 + 4NaOH \rightarrow CHI_3 + 3NaI + HCOONa + 3H_2O$$

$$CH_3COR + 3I_2 + 4NaOH \rightarrow CHI_3 + 3NaI + RCOONa + 3H_2O$$

A pale yellow precipitate of iodoform, CHI_3, is formed.

Ethanal is the only aldehyde to give this precipitate. All methyl ketones will also do this, thus pentan-2-one, $CH_3COCH_2CH_2CH_3$, will but pentan-3-one, $CH_3CH_2COCH_2CH_3$, will not.

Topic 17C Carboxylic acids and derivatives

Carboxylic acids

Carboxylic acids contain the $-COOH$ group, as shown in Figure 23.

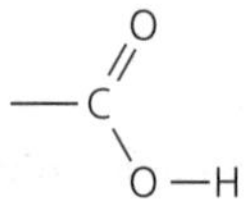

Figure 23 Structure of a carboxylic acid

Their general formula is $C_nH_{2n+1}COOH$, or RCOOH, where R is an alkyl group. They are all weak acids, and so they are only slightly ionised in water:

$$RCOOH + H_2O \rightarrow H_3O^+ + RCOO^-$$

$$K_a = \frac{[H_3O^+][RCOO^-]}{[RCOOH]}$$

Knowledge check 9

Why does the product of the reaction of methanal with hydrogen cyanide have no effect on plane-polarised light?

Exam tip

The lone pair of electrons on the δ– nitrogen attacks the δ carbon atom. Water is then lost and a double bond forms between the C and the N.

Exam tip

Ethanol (CH_3CH_2OH) and secondary alcohols containing th $CH_3CH(OH)$ group als undergo this reaction They are initially oxidised to ethanal and methyl ketones, respectively, which then react further to produce iodoform.

Vorked example 1

Worked example 1

Calculate the pH of a solution containing $0.123 \, mol \, dm^{-3}$ ethanoic acid ($K_a = 1.7 \times 10^{-5}$).

Answer

$$K_a = [H_3O^+] \times \frac{[CH_3COO^-]}{[CH_3COOH]}$$

$$[H_3O^+] = [CH_3COO^-]$$

$$[H_3O^+] = \sqrt{K_a \times [CH_3COOH]} = \sqrt{(1.7 \times 10^{-5} \times 0.123)} = 0.001446$$

$$pH = -\log 0.001446 = 2.84$$

Physical properties

The boiling temperature of carboxylic acids is higher than both aldehydes and alcohols with the same number of carbon atoms. Hydrogen bonding occurs between the $\delta+$ hydrogen of the OH group and the lone pair of electrons on the oxygen of the C=O group in another molecule, as shown in Figure 24. This is stronger than in alcohols as dimers are formed.

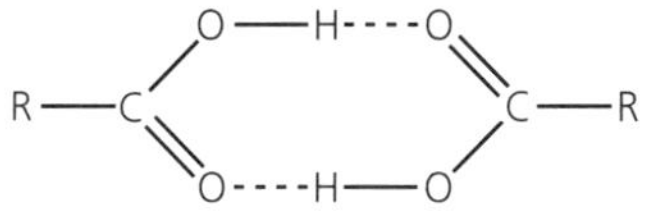

Figure 24 Hydrogen bonding in carboxylic acids

Acids are also able to form hydrogen bonds with water, so the lower members of the homologous series are water soluble.

Preparation

Oxidation of primary alcohols

When a primary alcohol (or an aldehyde) is heated under reflux with a solution of potassium dichromate(VI) in sulfuric acid, it is oxidised to a carboxylic acid:

$$CH_3CH_2CH_2OH + 2[O] \rightarrow CH_3CH_2COOH + H_2O$$

Hydrolysis of nitriles

When a nitrile is heated under reflux with a dilute acid such as hydrochloric acid, it is hydrolysed to a carboxylic acid:

$$CH_3CH_2CN + 2H_2O + HCl \rightarrow CH_3CH_2COOH + NH_4Cl$$

The hydroxynitrile, $CH_3CH(OH)CN$, formed from the reaction of ethanal with hydrogen cyanide can be hydrolysed to 2-hydroxypropanoic acid (lactic acid):

$$CH_3CH(OH)CN + 2H_2O + HCl \rightarrow CH_3CH(OH)COOH + NH_4Cl$$

Exam tip

Remember that the bond angle around the hydrogen-bonded hydrogen atom is 180°.

Exam tip

You must be able to draw labelled diagrams for this preparation and for the preparation of an aldehyde by oxidation of alcohols.

Reactions of carboxylic acids

Reduction with lithium aluminium hydride

Carboxylic acids are reduced to *primary* alcohols by lithium aluminium hydride. This is a two-stage process. The first stage is to add lithium aluminium hydride to a solution of the organic acid in dry ether. The second is to hydrolyse the adduct formed with aqueous acid. The reducing agent is represented by [H] in the overall equation:

$$CH_3COOH + 4[H] \rightarrow CH_3CH_2OH + H_2O$$

Reaction with bases

- Ethanoic acid reacts with aqueous sodium hydroxide to give a salt, sodium ethanoate:

$$CH_3COOH + NaOH \rightarrow CH_3COO^-Na^+ + H_2O$$

- It reacts with sodium carbonate (solid or in solution), giving off bubbles of carbon dioxide:

$$2CH_3COOH + Na_2CO_3 \rightarrow 2CH_3COO^-Na^+ + H_2O + CO_2$$

- It reacts with aqueous sodium hydrogen carbonate, giving off bubbles of carbon dioxide:

$$CH_3COOH + NaHCO_3 \rightarrow CH_3COO^-Na^+ + H_2O + CO_2$$

This reaction is a test for an acid.

Reaction with phosphorus(v) chloride

If phosphorus(v) chloride (phosphorus pentachloride) is added to a dry sample of ethanoic acid, steamy fumes of hydrogen chloride are given off and an acid chloride is produced. The organic product is ethanoyl chloride:

$$CH_3COOH + PCl_5 \rightarrow CH_3COCl + HCl + POCl_3$$

Reaction with alcohols: esterification

When ethanoic acid is warmed under reflux with an alcohol, in the presence of a few drops of concentrated sulfuric acid catalyst, an ester is formed. With ethanol, ethyl ethanoate is produced:

$$CH_3COOH + C_2H_5OH \rightleftharpoons CH_3COOC_2H_5 + H_2O$$

If the products of this reaction are poured into cold water, the characteristic sweet smell of an ester can be detected. Many simple esters are used as artificial food flavourings.

Acyl chlorides

Acyl chlorides contain the group shown in Figure 25.

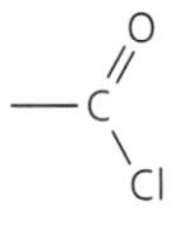

Figure 25 Structure of an acyl chloride

In the reaction betwee aqueous ethanoic acid and sodium hydrogencarbonate, is the sign of ΔS_{system} positive or negative? Explain your answer.

Exam tip

Alcohols (and water) also give steamy fum with phosphorus(v) chloride because this is a reaction of the –C group.

Exam tip

Do not write the formula of ethylethanoate as $CH_3OCOC_2H_5$ as that could also be the formula of the methyl ester of propanoic acid, whic is normally written a $C_2H_5COOCH_3$.

the carbon atom of the C=O group is δ+ (more so than in carboxylic acids). Therefore, acyl chlorides react with nucleophiles.

Reaction with water

Acyl chlorides are hydrolysed by water. A carboxylic acid is produced. For example:

$$CH_3COCl + H_2O \rightarrow CH_3COOH + HCl$$

Reaction with alcohols

On mixing, a rapid reaction takes place at room temperature and an ester is formed. For example, ethanoyl chloride reacts with propan-1-ol to form 1-propylethanoate:

$$CH_3COCl + CH_3CH_2CH_2OH \rightarrow CH_3COOCH_2CH_2CH_3 + HCl$$

Reaction with ammonia

With concentrated ammonia solution, an amide is rapidly formed. For example, with ethanoyl chloride the product is ethanamide:

$$CH_3COCl + 2NH_3 \rightarrow CH_3CONH_2 + NH_4Cl$$

Reaction with amines

In this reaction, a substituted amide is produced, which contains the –CONH– group. For example:

$$CH_3COCl + C_2H_5NH_2 \rightarrow CH_3CONHC_2H_5 + HCl$$

Esters

Esters contain the group shown in Figure 26.

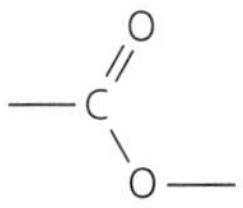

Figure 26 Structure of an ester

Hydrolysis with acids

Esters are *reversibly* hydrolysed when heated under reflux with an aqueous solution of a strong acid, which acts as a catalyst. Ethyl ethanoate is hydrolysed to ethanoic acid and ethanol:

$$CH_3COOC_2H_5 + H_2O \rightleftharpoons CH_3COOH + C_2H_5OH$$

Reaction with alkalis

When an ester is heated under reflux with an aqueous solution of an alkali such as sodium hydroxide, the ester is hydrolysed to an alcohol and the salt of a carboxylic acid in an *irreversible* reaction:

$$CH_3COOCH_3 + NaOH \rightarrow CH_3OH + CH_3COONa$$

methyl ethanoate   methanol   sodium ethanoate

> **Exam tip**
>
> This is a rapid reaction that goes to completion, unlike the slow, reversible esterification of a carboxylic acid with an alcohol.

> **Exam tip**
>
> If the product is then acidified, the carboxylic acid is formed.

Fats are esters of propane-1,2,3-triol (glycerol) and large carboxylic acids such as stearic acid ($C_{17}H_{35}COOH$). When heated under reflux with aqueous sodium hydroxide, sodium stearate ($C_{17}H_{35}COONa$) is produced. This is a soap and the reaction is called **saponification**, as shown in Figure 27.

$$CH_2OOCC_{17}H_{35}$$
$$|$$
$$CHOOCC_{17}H_{35} \ + \ 3NaOH \ \rightarrow \ CH_2(OH)CH(OH)CH_2OH \ + \ 3C_{17}H_{35}COONa$$
$$|$$
$$CH_2OOCC_{17}H_{35}$$

Figure 27 Saponification

Transesterification reactions

When an ester is reacted with an alcohol in the presence of an alkaline catalyst, a transesterification reaction takes place:

$$CH_3COOC_2H_5 \ + \ CH_3OH \ \rightleftharpoons \ CH_3COOCH_3 \ + \ C_2H_5OH$$

ethyl ethanoate methanol methyl ethanoate ethanol

A similar reaction takes place with a carboxylic acid:

$$CH_3COOC_2H_5 \ + \ HCOOH \ \rightleftharpoons \ HCOOC_2H_5 \ + \ CH_3COOH$$

ethyl ethanoate methanoic acid ethyl methanoate ethanoic acid

Use of this reaction is made in the manufacture of biodiesel and of low-fat spreads.

- Biodiesel — vegetable oils are esters of propane-1,2,3-triol (glycerol) and mainly unsaturated acids. They are too viscous to be a good fuel, so they are reacted with methanol in a transesterification reaction. The mixture of methyl esters is biodiesel:

  ester of propane-1,2,3-triol + $3CH_3OH \rightarrow$
          three esters of methanol + propane-1,2,3-triol

- Low-fat spreads — vegetable oils can be hardened to make margarine in two ways. One is partial hydrogenation, but this converts some of the isomers into the *trans* form, which is bad for health. The other alternative is to carry out a partial transesterification reaction with a saturated acid such as stearic acid, $C_{17}H_{35}COOH$. This replaces one of the unsaturated acids in the propane-1,2,3-ester and so increases the melting temperature.

Polyesters

These are formed by a *condensation* reaction between monomers with two functional groups. The most common is between a monomer with two COOH or two COCl groups and one with two OH groups.

Diacyl chlorides react with diols to form a condensation polymer:

$$n\text{ClOC}(CH_2)_4\text{COCl} + n\text{HOCH}_2CH_2CH_2OH \rightarrow \left[\!\!\begin{array}{c} \text{C}-(CH_2)_4-\text{C}-O-CH_2-CH_2O \\ \| \qquad\qquad\quad \| \\ O \qquad\qquad\quad O \end{array}\!\!\right]_n$$

> **Exam tip**
>
> The presence of doub[...] bonds makes packing less efficient. This results in weaker intermolecular force[...] and a lower melting temperature. Replac[...] unsaturated acid residues with saturat[...] ones ensures better packing and, therefor[...] a higher melting temperature.

> **Exam tip**
>
> A condensation reaction between tw[...] molecules results in the loss of water or some other simple molecule such as HC[...]

rylene is a polyester made from benzene-1,4-dicarboxylic acid and ethane-1,2-diol, shown in Figure 28.

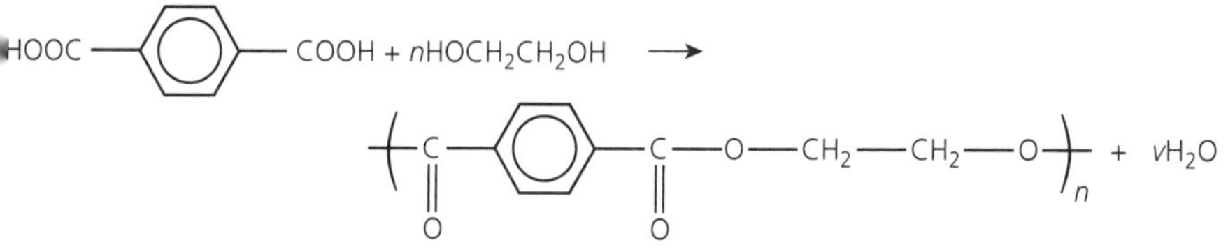
HOOC
COOH + nHOCH2CH2OH
C
O
C
O
CH2
CH2
O
n
+ vH2O

Figure 28 Structure of terylene

pol is a polyester which is biodegradable and is made from 3-hydroxybutanoic acid, $I_3CH(OH)CH_2COOH$. This monomer has one OH group and one COOH group l so can form a polyester with other 3-hydroxybutanoic acid molecules. Figure 29 ws a fragment showing *two* repeat units.

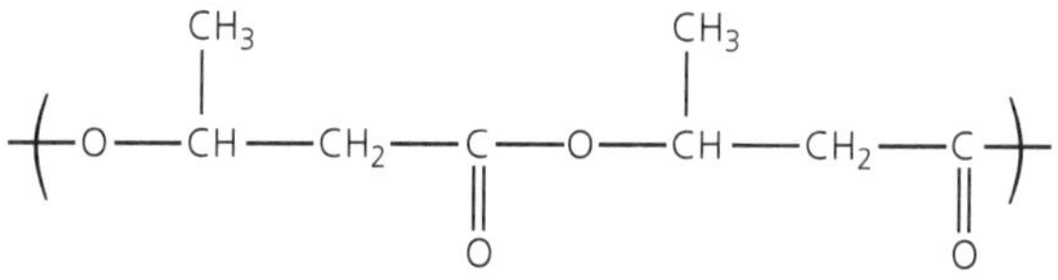
CH3
CH3
O
CH
CH2
C
O
CH
CH2
C
O
O

Figure 29 A fragment of biopol

Exam tip

Polyesters have a repeat unit with *four* oxygen atoms — one at each end and two in the middle.

ımmary

ter studying this topic, you should be able to:

- draw geometric and optical isomers and explain how they arise
- explain the physical properties of carbonyl compounds and carboxylic acids
- draw diagrams for the preparation of aldehydes and ketones from alcohols
- write equations and make observations for the reactions, if any, of aldehydes and ketones with 2,4-dinitrophenylhydrazine, alkaline iodine, Fehling's solution, Tollens' reagent and acidified dichromate(VI) ions
- draw the mechanism for the nucleophilic addition of HCN to a carbonyl compound and explain the effect of plane-polarised light on the product
- draw the mechanism for the nucleophilic substitution of HCN with halogenalkanes
- describe the preparation of carboxylic acids from alcohols, esters and nitriles
- write equations for the reactions of carboxylic acids with bases, carbonates, alcohols, PCl_5 and $LiAlH_4$ and for acyl chlorides with water, alcohol, ammonia and amines
- describe transesterification and explain its use in the production of biodiesel and low-fat spreads
- draw repeat units of polyesters

■ Topic 18 Organic chemistry III

Topic 18A Arenes

Benzene

Arenes are sometimes called aromatic compounds. They contain a benzene ring.

Benzene (C_6H_6) is a cyclic compound that has six carbon atoms in a hexagonal ring. Early theories suggested that there were alternate single and double bonds between the carbon atoms, but this did not fit with later experimental evidence. It was shown that all the carbon–carbon bonds are the same length and that the molecule is planar.

Two modern theories are used to explain the structure:

— The Kekulé version assumes that benzene is a **resonance hybrid** between the two structures, as shown in Figure 30.

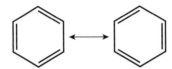

Figure 30 Kekulé theory of benzene structure

— The other theory assumes that each carbon atom is joined by a σ-bond to each of its two neighbours, and by a third σ-bond to a hydrogen atom. The fourth bonding electron is in a p-orbital, and the six p-orbitals overlap above and below the plane of the ring of carbon atoms. This produces a **delocalised** π-system of electrons, as shown in Figure 31.

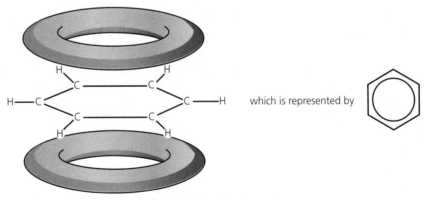

which is represented by

Figure 31 Delocalised p system

Thermochemical evidence
Via enthalpy of hydrogenation

Benzene is more stable than 'cyclohexatriene', which is the theoretical compound with three single and three localised double carbon–carbon bonds. The amount by which

...s stabilised can be calculated from the enthalpies of hydrogenation, is shown in ...gures 32 and 33.

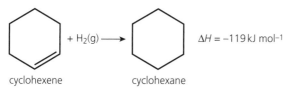

cyclohexene   cyclohexane

Figure 32 Enthalpies of cyclohexane

...erefore, ΔH for the addition to three localised double bonds in 'cyclohexatriene' ...uld be $3 \times -119 = -357$ kJ. However:

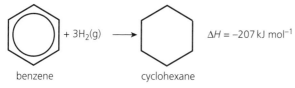

benzene   cyclohexane

Figure 33 Enthalpies of benzene

...us, 150 kJ *less* energy is given out because of benzene's unique structure. This is ...ed the **delocalisation stabilisation energy** or **resonance energy** and can be ...wn in the enthalpy-level diagram in Figure 34.

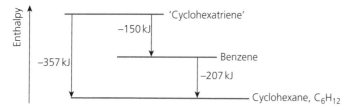

Figure 34 Enthalpy-level diagram showing delocalisation stabilisation energy

bond enthalpies

...amount by which benzene is stabilised can also be calculated from average bond ...halpies. The enthalpy of formation of gaseous benzene is $+83$ kJ mol^{-1}.

...value for the theoretical molecule 'cyclohexatriene' can be found using the Hess's ...cycle, shown in Figure 35.

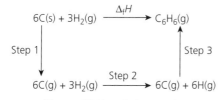

Figure 35 Hess's law cycle

1 equals $6 \times$ enthalpy of atomisation of carbon ($\Delta_a H$) $= 6 \times (+715) = +4290$ kJ

2 equals $3 \times$ H–H bond enthalpy $= 3 \times (+436) = +1308$ kJ

Exam tip

Cyclohexene has the same reactions as alkenes such as ethene.

Exam tip

Remember that bond breaking is endothermic and bond making exothermic.

Step 3 equals enthalpy change of bonds made:
- three C–C = $3 \times (-348) = -1044$ kJ
- three C=C = $3 \times (-612) = -1836$ kJ
- six C–H = $6 \times (-412) = -2472$ kJ

Total = -5352 kJ

The $\Delta_f H$ of 'cyclohexatriene' = $\Delta H_{step\ 1} + \Delta H_{step\ 2} + \Delta H_{step\ 3} = +4290 + 1308 + (-5352)$ $= +246$ kJ mol^{-1}. The actual enthalpy of formation of gaseous benzene is $+83$ kJ mol^{-1}. The value calculated above is 163 kJ more and equals the resonance energy of benzene.

Knowledge check

Why is the value of the resonance energy calculated from bond enthalpies slightly different to that calculated from enthalpy of hydrogenation?

X-ray diffraction evidence

X-ray diffraction shows the position of the centre of atoms. If the diffraction pattern of benzene is analysed, it clearly shows that all the bond lengths between the carbon atoms are the same, as shown in Table 8.

Table 8 Bond lengths between carbon atoms in benzene and cyclohexene

Bond	Bond length/nm
All six carbon–carbon bonds in benzene	0.14
Carbon–carbon single bond in cyclohexene	0.15
Carbon–carbon double bond in cyclohexene	0.13

Infrared evidence

Comparison of the infrared spectrum of aromatic compounds with those of aliphatic compounds containing a C=C group showed slight differences. The C–H stretching vibration in benzene is at 3036 cm^{-1} and the C=C stretching is at 1479 cm^{-1}, whereas the equivalent vibrations in an aliphatic compound such as cyclohexene are at 3023 cm^{-1} and 1438 cm^{-1}.

Reactions of benzene

Combustion

Benzene burns in a limited amount of air with a smoky flame. Combustion is incomplete and particles of carbon are formed.

Free radical addition

The double bond in benzene is not as susceptible to addition as is the double bond in alkenes. However, it does react with hydrogen in the presence of a hot nickel catalyst to form cyclohexane:

$$C_6H_6 + 3H_2 \rightarrow C_6H_{12}$$

It also reacts with bromine in the presence of intense ultra-violet light which splits the bromine molecules into atoms:

$$C_6H_6 + 3Br_2 \rightarrow C_6H_6Br_6$$

Electrophilic substitution

An **electrophile** is a species that attacks an electron-rich site.

Reaction with bromine: halogenation

Dry benzene reacts with liquid bromine in the presence of iron (or a catalyst of anhydrous iron(III) bromide). Steamy fumes of hydrogen bromide are given off and bromobenzene (C_6H_5Br) is formed, as shown in Figure 36.

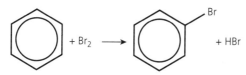

Figure 36 The reaction of benzene with liquid bromine to produce hydrogen bromide and bromobenzene

The mechanism for this reaction is as follows. The catalyst, anhydrous iron(III) bromide, is made by the reaction of iron with bromine:

$$Fe + 1\tfrac{1}{2}Br_2 \rightarrow FeBr_3$$

This then reacts with more bromine, forming the electrophile Br^+:

$$Br_2 + FeBr_3 \rightarrow Br^+ + [FeBr_4]^-$$

The Br^+ attacks the π-electrons in the benzene ring, forming an intermediate with a positive charge. Finally, the $[FeBr_4]^-$ ion removes an H^+ from benzene, producing hydrogen bromide (HBr) and re-forming the catalyst ($FeBr_3$) as shown in Figure 37.

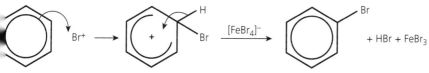

Figure 37 The addition of Br^+ to benzene, producing hydrogen bromide and reforming $FeBr_3$

Comparison of the reaction of bromine with alkenes

Benzene reacts with bromine in an electrophilic *substitution* reaction, whereas alkenes react in an electrophilic *addition* reaction. The addition of Br^+ to benzene is similar to the first step of the addition of bromine to ethene. The difference arises at the next step. The benzene intermediate loses an H^+, thus regaining *the stability of the delocalised π-system*, whereas the intermediate with ethene adds a Br^- ion. A catalyst must be present for the addition of Br^+ to benzene, because the activation energy of the first step is higher than that for the addition to ethene. No catalyst is needed for the addition of bromine to alkenes.

Reaction with nitric acid: nitration

When benzene is warmed with a mixture of concentrated nitric and sulfuric acids, a nitro-group (NO_2) replaces a hydrogen atom in the benzene ring. Nitrobenzene and water are produced, as shown in Figure 38.

Exam tip

A catalyst is *always* needed, since the stability due to delocalisation causes the reactions to have large activation energies.

Exam tip

Do *not* state that the catalyst is iron. It is iron(III) bromide which is formed by the reaction of iron with bromine.

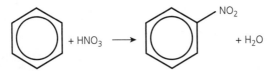

Figure 38 The reaction of benzene with nitric acid to produce nitrobenzene and water

The sulfuric acid reacts with the nitric acid to form the electrophile NO_2^+. The temperature must not go above 50°C or some dinitrobenzene $(C_6H_4(NO_2)_2)$ is formed.

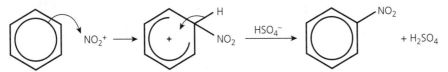

Figure 39 The reaction of benzene with the electrophile NO_2 producing nitrobenzene and sulfuric acid

Friedel–Crafts reactions

Reaction with halogenoalkanes

In the presence of an anhydrous aluminium chloride catalyst, alkyl groups (e.g. C_2H_5) can be substituted into the ring. In the reaction between benzene and iodoethane, the products are ethylbenzene and hydrogen iodide, as shown in Figure 40.

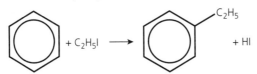

Figure 40 The reaction of benzene with iodoethane producing ethylbenzene and hydrogen iodide

The reaction mixture must be *dry*.

The electrophile is produced by the reaction of the catalyst with the halogenoalkane:

$$CH_3CH_2I + AlCl_3 \rightarrow CH_3CH_2^+ + [AlCl_3I]^-$$

The positive carbon atom attacks the π-system in the benzene ring, as shown in Figure 41.

Step 1: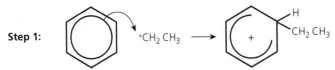

Figure 41 The positive carbon atom from iodoethane attacking the π-system in the benzene ring

The intermediate loses a H^+ ion so as to *regain* the stability of the benzene ring, as shown in Figure 42.

Step 2:

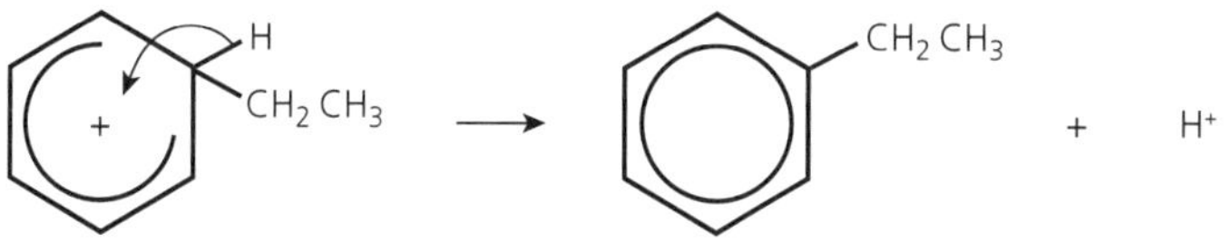

igure 42 The intermediate losing a H⁺ ion and regaining stability of the benzene ring

e catalyst is regenerated by the reaction:

$H^+ + [AlCl_3I]^- \rightarrow HI + AlCl_3$

action with acyl chlorides

the presence of an anhydrous aluminium chloride catalyst, benzene reacts with
l chlorides to form ketones. In the reaction between benzene and ethanoyl
loride, the products are phenylethanone and hydrogen chloride, as shown in
gure 43.

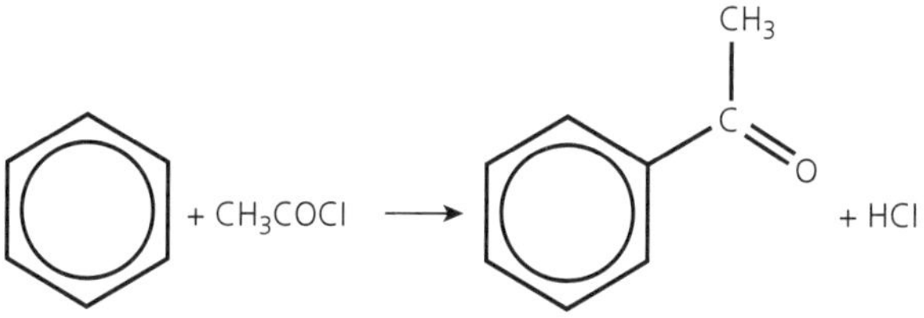

Figure 43 The reaction between benzene and ethanoyl chloride, producing
phenylethanone and hydrogen chloride

e electrophile is produced by the reaction of the acyl chloride with the catalyst:

$CH_3COCl + AlCl_3 \rightarrow CH_3C^+O + [AlCl_4]^-$

e positive carbon atom attacks the π-system in the benzene ring, as shown in
re 44.

Step 1:

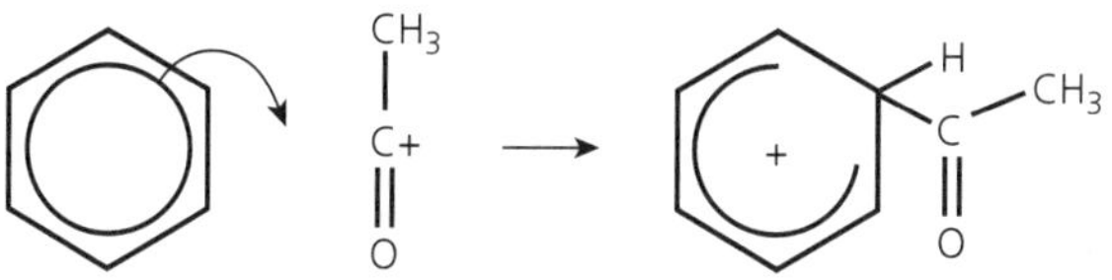

ure 44 The positive carbon atom from acyl chloride attacking the n-system in the
benzene ring

intermediate loses a H⁺ ion so as to *regain* the stability of the benzene ring, as
vn in Figure 45.

Step 2:

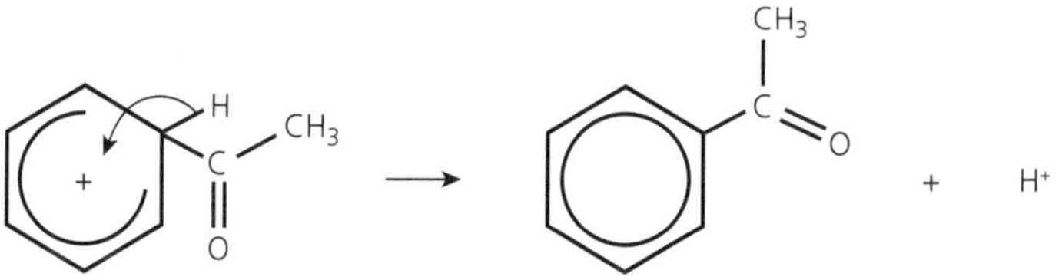

ure 45 The intermediate losing a H⁺ ion and regaining stability of the benzene ring

Exam tip

The reagents must be
dry and the aluminium
chloride anhydrous.

The catalyst is regenerated by the reaction:

$$H^+ + [AlCl_4]^- \rightarrow HCl + AlCl_3$$

Exam tip

Make sure that:

- the curly arrow starts from the delocalised ring
- the intermediate has a delocalised ring around five carbon atoms and is positively charged
- the curly arrow starts from the σ-bond between a carbon and the hydrogen in the final step

Phenol

Phenol (C_6H_5OH) contains an –OH group on a benzene ring. The p_z lone pair of electrons on the oxygen atom becomes part of the delocalised π-system and makes phenol more susceptible to attack by electrophiles.

Reaction with bromine

The delocalised ring in phenol is so electron-rich that it is attacked by bromine water, in an electrophilic substitution reaction, as shown in Figure 46. The brown bromine water is decolorised and a white precipitate of 2,4,6-tribromophenol and a solution of hydrogen bromide are formed. No catalyst is needed.

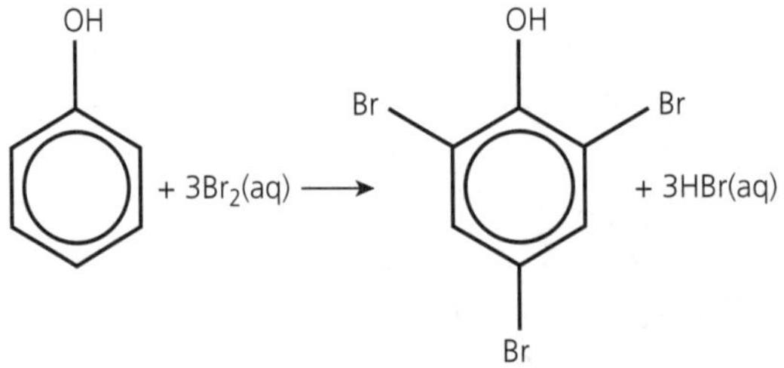

Figure 46 Reaction of phenol and bromine water, producing 2,4,6-tribromophenol and hydrogen bromide

Knowledge check

Explain why methylbenzene react faster than benzene in electrophilic substitution.

Exam tip

Remember that the three Br atoms each replace a hydrogen atom in the ring, so there must be 3HBr on the right of the equation.

Summary

After studying this topic, you should be able to:

- define the term electrophile
- estimate resonance energy of benzene from hydrogenation and bond enthalpy data
- write equations and state conditions for the reactions of benzene with bromine, nitric acid and the Friedel–Crafts reactions
- draw mechanisms for the halogenation, nitration and Friedel–Crafts reactions of benzene
- explain why the ring in methylbenzene is slightly activated and that in phenol very activated and why phenol reacts with bromir in the absence of a catalyst

Topic 18B Amines, amides, amino acids and proteins

The types of organic nitrogen compounds that are included are:

amines: these contain a nitrogen atom covalently bonded to a carbon atom, usually in a NH_2 group, as in 1-butylamine ($CH_3CH_2CH_2CH_2NH_2$)

amides: these contain the group shown in Figure 47

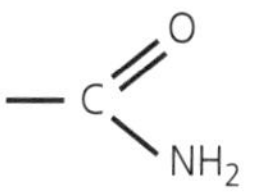

Figure 47 Group present in amides

condensation polymers

amino acids: these contain a NH_2 and a COOH group, as in aminoethanoic acid (NH_2CH_2COOH), which is also called glycine

Amines

Primary amines contain the $-NH_2$ group — for example, ethylamine ($C_2H_5NH_2$).

Secondary amines contain the $-NH$ group — for example, diethylamine (($C_2H_5)_2NH$).

Tertiary amines contain the $-N<$ group — for example, triethylamine (($C_2H_5)_3N$).

Exam tip

All amines contain a lone pair of electrons on the nitrogen atom, which is $\delta-$.

Physical properties

Amines have a fish-like smell.

Primary and secondary amines form intermolecular hydrogen bonds between the lone pair of electrons on the nitrogen in one molecule and the $\delta+$ hydrogen in another molecule.

All amines form hydrogen bonds with water. The lone pair of electrons on the $\delta-$ nitrogen forms a hydrogen bond with the $\delta+$ hydrogen in water. This means that the members of the homologous series of amines of low molar mass are miscible with water. Others, such as phenylamine ($C_6H_5NH_2$) are partially soluble.

Knowledge check 13

Explain why ethylamine has a higher boiling temperature than propane.

Preparation

From halogenoalkanes

A halogenoalkane is left to stand with excess concentrated ammonia, a primary amine is formed.

$$C_2H_5I + 2NH_3 \rightarrow C_2H_5NH_2 + NH_4I$$

From nitriles

A nitrile, such as 1-butanenitrile, $CH_3(CH_2)_2C\equiv N$, can be reduced to an amine by reduction to 1-butylamine with lithium tetrahydridoaluminate (lithium aluminium hydride) in ether followed by hydrolysis using dilute acid.

$$CH_3(CH_2)_2C\equiv N + 4[H] \rightarrow CH_3(CH_2)_2CH_2NH_2$$

By reduction of nitrocompounds

Aromatic nitro compounds such as nitrobenzene, $C_6H_5NO_2$, can be reduced to aromatic amines by concentrated hydrochloric acid and tin.

If nitrobenzene is heated under reflux with tin and concentrated hydrochloric acid, the NO_2 group is reduced to an $NH_3^+Cl^-$ group. Phenylamine is set free on addition of alkali. The overall reaction is:

$$C_6H_5NO_2 + 6[H] \rightarrow C_6H_5NH_2 + 2H_2O$$

The phenylamine is removed from the reaction mixture by steam distillation (see page 55).

Reactions

With water

Amines are weak bases. The lone pair of electrons on the nitrogen atom can form a bond with the $\delta+$ hydrogen in water and so deprotonate water in a reversible reaction:

$$CH_3(CH_2)_2CH_2NH_2 + H_2O \rightleftharpoons CH_3(CH_2)_2CH_2NH_3^+ + OH^-$$

As OH^- ions are formed, the solution is alkaline (pH > 7).

This is similar to the reaction of ammonia and water:

$$NH_3 + H_2O \rightleftharpoons NH_4^+ + OH^-$$

1-butylamine is a stronger base than ammonia because the alkyl groups are slightly electron pushing and so the nitrogen atom becomes more $\delta-$. Phenylamine is a weaker base than ammonia because the lone pair of electrons on the nitrogen atom become partially incorporated in the benzene ring's π system.

With acids

Amines are weak bases, just like ammonia, and so they react with both strong and weak acids to form salts. With hydrochloric acid, the product is 1-butylammonium chloride:

$$CH_3(CH_2)_2CH_2NH_2 + HCl \rightarrow CH_3(CH_2)_2CH_2NH_3^+Cl^-$$

This is similar to the reaction of ammonia with hydrochloric acid:

$$NH_3 + HCl \rightarrow NH_4^+Cl^-$$

With a weak acid, such as ethanoic acid, the product is 1-butylammonium ethanoate:

$$CH_3CH_2CH_2CH_2NH_2 + CH_3COOH \rightarrow CH_3CH_2CH_2CH_2NH_3^+CH_3COO^-$$

With acyl chlorides

Amines react rapidly with acyl chlorides to form a substituted amide:

$$CH_3(CH_2)_2CH_2NH_2 + CH_3COCl \rightarrow CH_3CONHCH_2(CH_2)_2CH_3 + HCl$$

The organic product is called N-butylethanamide. The N-butyl part of the name shows that the butyl group is attached to the *nitrogen* atom in the amide.

With halogenoalkanes

Primary amines react with halogenoalkanes when warmed in ethanolic solution to form the salt of a secondary amine:

Knowledge check

Ethylamine smells of fish. When acid is added the smell disappears and when alkali is then added the smell returns. Explain these observations.

$$C_2H_5NH_2 + C_2H_5I \rightarrow (C_2H_5)_2N^+H_2I^-$$

he secondary amine can be regenerated by adding alkali:

$$(C_2H_5)_2N^+H_2I^- + OH^- \rightarrow (C_2H_5)_2NH + I^- + H_2O$$

/ith excess halogenoalkane a tertiary or even a quaternary amine salt is formed.

$$(C_2H_5)_2N^+H_2I^- \xrightarrow{C_2H_5I} (C_2H_5)_3N^+HI^- \xrightarrow{C_2H_5I} (C_2H_5)_4N^+I^-$$

Vith copper(II) ions

liphatic amines form complex ions similar to that formed by ammonia:

$$[Cu(H_2O)_6]^{2+} + 4C_2H_5NH_2 \rightarrow [Cu(C_2H_5NH_2)_4(H_2O)_2]^{2+} + 4H_2O$$

$$[Cu(H_2O)_6]^{2+} + 4NH_3 \rightarrow [Cu(NH_3)_4(H_2O)_2]^{2+} + 4H_2O$$

he ethylamine reacts with hydrated copper(II) ions in a ligand-exchange reaction to
rm a purple-blue solution.

Exam tip

1,2-diaminoethane is a
bidentate ligand and
forms a purple solution
of $[Cu(H_2NCH_2CH_2NH_2)_2(H_2O)_2]^{2+}$ ions.

mides

mides are prepared by the reaction of an acyl chloride with ammonia:

$$CH_3COCl + NH_3 \rightarrow CH_3CONH_2 + HCl$$

nanoyl chloride   ethanamide

nides are hydrolysed by heating with dilute hydrochloric acid or aqueous alkali:

$$CH_3CONH_2 + HCl + H_2O \rightarrow CH_3COOH + NH_4Cl$$

$$CH_3CONH_2 + NaOH \rightarrow CH_3COO^-Na^+ + NH_3$$

ondensation polymers

ndensation polymers are formed when two compounds, each with two functional
ups, polymerise. Water, or another simple molecule, is eliminated every time two
lecules combine.

olyesters

ese are formed by a condensation reaction. The most common is between a
arboxylic acid or a diacyl chloride and a diol (see page 32).

olyamides

ese are formed from a monomer containing two NH_2 groups and another
taining two carboxylic acid or acyl chloride groups.

on is a polymer of 1,6-diaminohexane, $H_2N(CH_2)_6NH_2$, and hexane-1,6-dioyl
oride, $ClOC(CH_2)_4COCl$, with the elimination of HCl. The structure of two
eat units is shown in Figure 48.

Knowledge check 15

The biodegradable
polyester Biopol
is made from
3-hydroxybutanoic acid.
Draw a repeat unit of
this polymer.

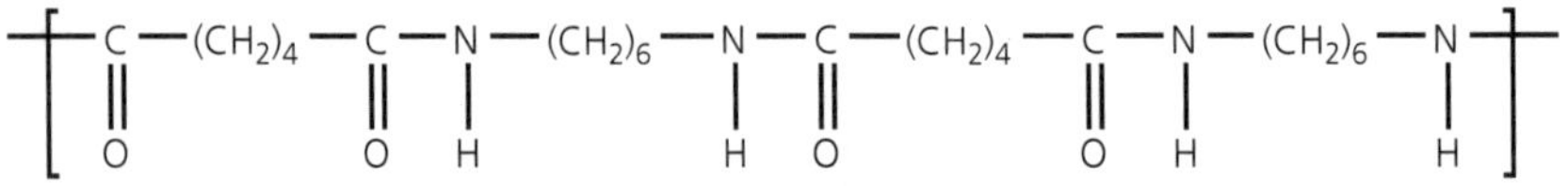

A polyamide

Figure 48 Two repeat units of the monomers within nylon

Kevlar® is a polymer of 1,4-diaminobenzene, $H_2NC_6H_4NH_2$, and benzene-1,4-dicarbonyl dichloride, $ClOCC_6H_4COCl$. One repeat unit is shown in Figure 49.

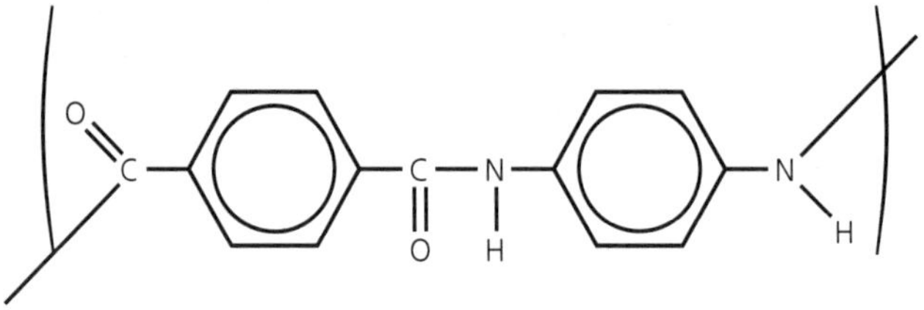

Figure 49 A repeat unit of the polymer Kevlar®

Polyamides have a higher melting point than most polymers. This is because of hydrogen bonding between the $\delta+$ hydrogen of the NH group in one chain with the $\delta-$ oxygen in the C=O group of another chain.

Amino acids

Amino acids contain both $-NH_2$ and $-COOH$ groups. They are found naturally in proteins. All except glycine, H_2NCH_2COOH, contain a chiral centre. For example, 2-aminopropanoic acid, $H_2NCH(CH_3)COOH$, has two optical isomers, as shown in Figure 50.

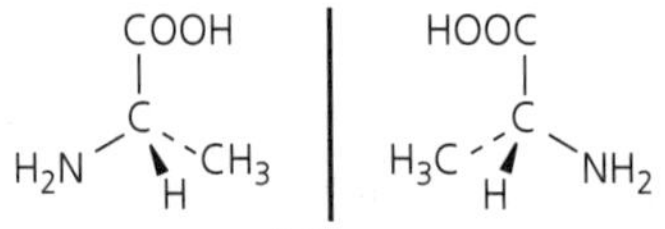

Mirror

Figure 50 Optical isomers of the amino acid 2-aminopropanoic acid

Physical properties

Solubility

Amino acids are water-soluble solids. The reason for this is that the acidic $-COOH$ group protonates the basic $-NH_2$ group, forming a **zwitterion**, which has a positive charge on one end and a negative charge on the other:

$$NH_2CH_2COOH \rightleftharpoons {}^+NH_3CH_2COO^-$$

The position of this equilibrium lies well to the right, so there are few neutral molecules in the solution. The ion–dipole attractions involving the $\delta+$ H and $\delta-$ O atoms in the water give rise to its solubility.

Exam tip

Make sure that you draw the amide –CONH– group in full when drawing the repeat unit of a polyamide.

Exam tip

Note that the repeat unit in all polyamides has two oxygen and two nitrogen atoms and the repeat unit in all polyesters has four oxygen atoms.

Exam tip

2-aminopropanoic acid is also called alanine (not to be confused with aniline, $C_6H_5NH_2$).

H

lycine and alanine are examples of neutral amino acids as the alkaline NH_2 group
cancelled because it is protonated by the acidic COOH group. However some, such
aspartic acid, $HOOCCH_2CH(NH_2)COOH$, have two COOH groups and so are
idic. Others, such as lysine, $NH_2(CH_2)_4CH(NH_2COOH$, have two NH_2 groups
d are alkaline.

ffect on plane-polarised light

ost amino acids are chiral. Natural amino acids are single enantiomers of the
rmula $H_2NCHRCOOH$. Natural alanine, $H_2NCH(CH_3)COOH$, is chiral and
tates the plane of polarisation of plane-polarised light. Glycine, H_2NCH_2COOH, is
t chiral and so has no effect on plane-polarised light.

elting point

ne ion–ion attractions between the *different* zwitterions result in the substance being
olid with a high melting point.

action with acids

ne $-NH_2$ group becomes protonated:

$$NH_2CH_2COOH + H^+ \rightarrow {}^+NH_3CH_2COOH$$

action with bases

e $-COOH$ group protonates the base:

$$NH_2CH_2COOH + OH^- \rightarrow NH_2CH_2COO^- + H_2O$$

action with ninhydrin

nhydrin reacts on heating with amino acids to form a deep-blue/purple colour.
is reaction is used to identify the positions of different amino acids after
romatographic separation — see pages 46 and 63.

oteins

teins contain a sequence of amino acids joined by a peptide bond. This is the same
d that joins the monomers in a polyamide. If two amino acids join, two isomeric
eptides are possible. When glycine and alanine react, two possible dipeptides are
ned, shown in Figure 51.

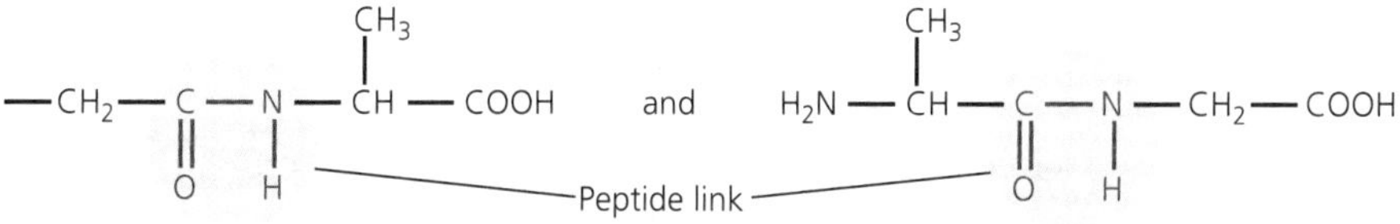

Figure 51 Isomeric dipeptides formed by the reaction of glycine and alanine

e peptide link $-CONH-$ is circled in these two structures.

teins are polypeptides. Insulin has 17 different amino acids and a total of 51 amino
d molecules joined by 50 peptide bonds. This is an example of **condensation
ymerisation**.

Exam tip

The reactions of amino
acids can be written
with the zwitterion
on the left, e.g.
$^+NH_3CH_2COO^- + H^+ \rightarrow$
$^+NH_3CH_2COOH$.

Knowledge check 16

Write the equation
for the reaction of the
zwitterion of glycine
with OH^- ions.

Thin-layer chromatography

The peptide links can be hydrolysed by prolonged heating with hydrochloric acid. This breaks the polypeptide down into its constituent amino acids, which can be separated by chromatography, such as thin-layer chromatography.

In thin-layer chromatography (TLC), as with all types of chromatography, there is a stationary phase and a moving phase. In TLC, the stationary phase is either silica gel or aluminium oxide immobilised on a flat inert sheet, which is usually made from glass or plastic.

- The mixture of amino acids is dissolved in a suitable solvent and a spot of the solution placed about 2 cm from the bottom of the plate.
- Spots of dissolved known amino acids are placed on the same plate at the same level.
- The plate is then dipped in a suitable eluent (the mobile phase) with the spots above the level of the liquid eluent and is placed in a sealed container. The eluent is drawn up the plate by capillary action.
- The plate is left until the eluent rises to the top of the plate.
- The plate is removed, sprayed with a solution of ninhydrin and heated.
- The ninhydrin reacts with the amino acids, producing a blue-purple colour.
- The height that the unknown has reached is compared with the heights reached by the known amino acids. Spots at the same height are caused by the same amino acid. This enables the amino acids in the unknown to be identified.

Knowledge check 1

Draw the structural formula of the *two* dipeptides that can be formed from glycine, NH_2CH_2COOH and valine, $(CH_3)_2CHCH(NH_2)COO$

Summary

After studying this topic, you should be able to:
- explain the physical properties of amines and amino acids
- write equations for the reactions of amines with water, acids, acyl chlorides, halogenoalkanes and copper(ii) ions
- describe the preparation of phenylamine from nitrobenzene

- describe condensation polymerisation, and draw the repeat units involved
- recall the reactions of amino acids with H^+ and OH^- ions, and with ninhydrin
- understand the formation of peptide links in proteins

Functional-group analysis

You must give the *full* name or formula of the reagents used in any test. If a colour change is observed, then you must state the colour *before* and *after* the test.

Alkenes

The functional group is C=C. Add bromine dissolved in water. The brown solution becomes colourless.

Halogenoalkanes

Warm with a little aqueous sodium hydroxide mixed with ethanol. Then acidify with dilute nitric acid and add silver nitrate solution.

The observation depends on the type of halogenoalkane:
- Chloroalkanes give a white precipitate soluble in dilute ammonia solution.

Exam tip

Do *not* state that the bromine water goes clear. Clear solutions can be coloured.

Bromoalkanes give a cream precipitate insoluble in dilute ammonia solution but soluble in concentrated ammonia.

Iodoalkanes give a yellow precipitate, insoluble in concentrated ammonia.

Hydroxyl groups in acids and alcohols

Add phosphorus pentachloride to the dry compound. Steamy fumes are given off.

Acids

Add aqueous sodium hydrogencarbonate. A gas is evolved that turns limewater cloudy.

Alcohols

Add ethanoic acid and a few drops of concentrated sulfuric acid and warm gently. Pour into a beaker of cold water and carefully smell the product. A fruity or glue-like smell confirms an alcohol.

To distinguish primary and secondary alcohols from tertiary alcohols, warm with potassium dichromate(VI) in dilute sulfuric acid:
- Primary and secondary alcohols turn the orange solution green.
- Tertiary alcohols do not react.

To distinguish between primary and secondary alcohols, test as above and distil the product as it forms into ammoniacal silver nitrate solution.
- Primary alcohols form a silver mirror.
- Secondary alcohols do not react.

Carbonyl compounds

Add 2,4-dinitrophenylhydrazine solution (Brady's reagent). A red-orange precipitate is produced.

To distinguish between aldehydes and ketones, add Tollens' reagent and warm.
- Aldehydes form a silver mirror.
- Ketones do not react.

Alternatively, add Fehling's (or Benedict's) solution and warm:
- Aldehydes produce a red-brown precipitate.
- Ketones do not react, so the blue solution remains.

Iodoform reaction

Add the test compound to a mixture of iodine and aqueous sodium hydroxide. A pale-yellow precipitate of iodoform, CHI_3, is produced with the following:
- methyl ketones, because they contain the CH_3CO group
- secondary alcohols that contain the $CH_3CH(OH)$ group, which is oxidised by the iodine to a methyl ketone
- ethanal (CH_3CHO) and ethanol (CH_3CH_2OH)

Knowledge check 18

A substance with molecular formula $C_4H_6O_2$ decolourised bromine water, gave steamy fumes with PCl_5, a red precipitate with 2,4-dinitrophenylhydrazine and a silver mirror with Tollens' reagent. Draw its structural formula.

Core practical 15

Analysis of organic and inorganic compounds
Questions will be asked, particularly in paper 3, on the analysis of unknowns.

An organic unknown
An organic compound X gave a yellow precipitate with 2,4-dinitrophenyl hydrazine, did not change the colour of Fehling's solution but gave a pale yellow precipitate with iodine in alkali. The precipitate with 2,4-dinitrophenyl hydrazine was purified by recrystallisation and its melting point was found to be 117°C.

Use the data on page 28 and the information above to suggest the structural formula of compound X.

Answer
- The precipitate with 2,4-dintrophenylhydrazine shows that X is a carbonyl compound.
- As it did not change the colour of Fehling's solution, it is not an aldehyde, so it is a ketone.
- The precipitate of iodoform shows that it has a $CH_3C=O$ group.
- The melting point is consistent with X being butanone.

The structural formula of X is $CH_3COCH_2CH_3$.

An inorganic unknown
An inorganic compound Z is green. A solution of Z was tested as below. Comment on the result of each test and finally write the formula of Z.
- On addition of aqueous sodium hydroxide a green precipitate was obtained which did not dissolve in excess.
- On addition of aqueous ammonia a green precipitate was obtained which dissolved in excess to form a pale blue solution.
- When a solution of aqueous silver nitrate was added, a precipitate formed. This precipitate was filtered and treated first with dilute ammonia and then with concentrated ammonia. The precipitate did not dissolve in dilute ammonia but did in concentrated ammonia.

Answer
The three green hydroxides are those of iron(II), chromium(III) and nickel. Chromium(III) hydroxide is amphoteric and would dissolve in excess sodium hydroxide, so Z is either an iron(II) or a nickel salt.

Iron(II) hydroxide would not dissolve in excess ammonia, so Z is a nickel salt.

The precipitate was a silver halide. As it was insoluble in dilute ammonia it was not a precipitate of silver chloride. Its solubility in concentrated ammonia shows that the precipitate was silver bromide.

The formula of Z is $NiBr_2$.

Topic 18C Organic synthesis

Deduction of empirical and molecular formulae

The first step is finding the percentage composition of the substance. This is done by burning a known mass of the substance in excess air. The water is absorbed in a suitable drying agent, such as silica gel, and the carbon dioxide is absorbed by dry calcium oxide. This measures the masses of water and carbon dioxide formed. The route is then:
- mass of water → mass of hydrogen → % hydrogen
- mass of carbon dioxide → mass of carbon → % carbon
- 100 − (% of hydrogen + % of carbon) = % oxygen

Worked example 1

An organic compound X contains carbon, hydrogen and oxygen only. When 2.00 g of X was burnt in excess air, 1.46 g of water and 3.57 g of carbon dioxide were produced. Calculate the percentage composition of the elements in compound X.

Answer

1.46 g of H_2O contains $1.46 \times \frac{2}{18} = 0.162$ g hydrogen

% hydrogen $= 0.162 \times \frac{100}{2.00} = 8.1\%$

3.57 g of CO_2 contains $3.57 \times \frac{12}{44} = 0.9736$ g carbon

% carbon $= 0.9736 \times \frac{100}{2.00} = 48.7\%$

% oxygen $= 100 - 8.1 - 48.7 = 43.2\%$

The second step is to deduce the empirical formula from the percentage composition. The route for this is as follows:
  Divide each percentage by the relative atomic mass of the element.
  Divide by the smallest and round to one decimal place.
  If the answer is not a whole number ratio, multiply all by two (and if this still is not, try multiplying by three).

Worked example 2

A compound Z, which contained carbon, hydrogen and oxygen only, was found to contain 53.3% carbon and 11.1% hydrogen. Calculate its empirical formula.

Answer

Element	%	% ÷ A_r	÷ smallest
Carbon	53.3	53.3/12 = 4.4	4.4/2.2 = 2.0
Hydrogen	11.1	11.1/1 = 11.1	11.1/2.2 = 5.0
Oxygen	100 − 53.3 − 11.1 = 35.6	35.6/16 = 2.2	2.2/2.2 = 1.0

The empirical formula of compound Z is C_2H_5O.

The final step is to work out the molecular formula. For this, the relative molecular mass must be known. This will either be given or deduced from the mass spectrum, where the largest *m/z* value in the spectrum is assumed to be that of the molecular ion.

Worked example 3

The mass spectrum of compound Z has a peak due to the molecular ion at *m/z* = 90. Use your results from Worked example 2 to deduce its molecular formula.

Answer

The relative empirical mass is 45 which is half that of the *m/z* of the molecular ion, so the molecular formula of Z is twice the empirical formula and is $C_4H_{10}O_2$.

Reaction schemes

You are expected to be able to deduce methods of converting one organic substance into another. Such questions may involve a reaction scheme of up to four steps. You are required to identify:

- the reagent (by name or formula) for each step
- the conditions (if asked for)
- each intermediate in the synthesis

Common reactions are listed in Table 9. Those marked * cause the carbon chain to be lengthened; that marked # causes the carbon chain to be shortened.

> **Exam tip**
>
> Cover the right hand column of the table below with a sheet of paper, then see if you can remember how to make each conversion.

> **Exam tip**
>
> If you give *both* name and formula, both must be correct.

Table 9

Conversion	Reagents and conditions
Alkene to a halogenoalkane (addition)	Mix with gaseous hydrogen halide, e.g. hydrogen bromide, HBr
Alkene to a halogenoalcohol (addition)	Bubble into bromine water
Halogenoalkane to an alcohol (nucleophilic substitution)	Heat under reflux with aqueous sodium hydroxide
Halogenoalkane to an amine (nucleophilic substitution)	React with concentrated ammonia in aqueous alcohol
Halogenoalkane to an alkene (elimination)	Heat under reflux with potassium hydroxide in alcohol
Alcohol to a halogenoalkane (substitution)	Add phosphorus(v) chloride or potassium bromide and 50% sulfuric acid or moist red phosphorus and iodine
Primary alcohol to an aldehyde (partial oxidation)	Heat with potassium dichromate(vi) in dilute sulfuric acid and distil off the aldehyde as it forms
Primary alcohol to a carboxylic acid (complete oxidation)	Heat under reflux with potassium dichromate(vi) in dilute sulfuric acid
Secondary alcohol to a ketone (oxidation)	Heat under reflux with potassium dichromate(vi) in dilute sulfuric acid
*Aldehyde or ketone to a hydroxynitrile (nucleophilic addition)	Add hydrogen cyanide in a buffer solution of pH = 7
#Methyl ketone (e.g. $RCOCH_3$) or a secondary 2-ol (e.g. $RCH(OH)CH_3$ to the carboxylic acid RCOOH (iodoform reaction)	Add iodine mixed with aqueous sodium hydroxide
Nitrile (or hydroxynitrile) to a carboxylic acid (hydrolysis)	Heat under reflux with dilute sulfuric acid
Nitrile to a primary amine (reduction)	Lithium aluminium hydride in dry ether
*Benzene to an alkyl benzene or a ketone (Friedel–Crafts reaction)	React with a halogenoalkane or acyl chloride in presence of anhydrous aluminium chloride catalyst
Benzene to nitrobenzene	React with a mixture of concentrated nitric and sulfur acids at 50°C
Nitrobenzene to phenylamine (reduction)	Heat with tin and concentrated hydrochloride acids, then add alkali and steam distil off the product

Planning a synthesis

is assumed that the reactions of compounds with more than one functional group e the same as those of simple compounds each with one functional group. For ample, the reactions of $CH_2=CHCH_2CH(OH)CHO$ are those of an alkene, a condary alcohol and an aldehyde.

ometimes the conversion may require the carbon chain length to be altered, and this n be an important clue to the route. This means that one step in the reaction is the dition of hydrogen cyanide to a carbonyl compound, substitution by cyanide ions th a halogenoalkane, via a Grignard reagent (see below) or a Friedel–Crafts reaction th an arene. If the length of the carbon chain is decreased, the iodoform reaction ll be part of the synthesis.

ote that the carbon chain is the number of carbon atoms joined to each other. It is t necessarily the total number of carbon atoms in the molecule. For instance, the er ethyl propanoate ($CH_3CH_2COOC_2H_5$) has two carbon chains, one with three rbon atoms (propanoate) and one with two (ethyl).

the question asks how ethanoic acid can be prepared from ethene, ask yourself the lowing questions:

How can an acid be prepared?

swer: by oxidation of a primary alcohol

How can a primary alcohol be prepared?

swer: by hydrolysis of a halogenoalkane

Can a halogenoalkane be prepared from an alkene?

swer: yes, by addition of a hydrogen halide. The route is:

ethene $\xrightarrow{\text{HBr(g)}}$ bromoethane $\xrightarrow{\text{NaOH(aq)}}$ ethanol $\xrightarrow{\text{H}^+/\text{Cr}_2\text{O}_7^{2-}}$ ethanoic acid

Grignard reagents

se compounds also increase the length of the carbon chain. They consist of a rocarbon residue, R, joined to a magnesium atom that is bonded to a halogen, ally bromine or iodine. An example is methyl magnesium iodide, CH_3MgI. The on atom attached to the magnesium is very $\delta-$ and so is a powerful nucleophile attacks electron-deficient ($\delta+$) sites.

Grignard reagents are prepared in ether solution and all water must be excluded. y react with carbon dioxide and carbonyl compounds.

Carbon dioxide → carboxylic acids

en carbon dioxide is bubbled into a solution of a Grignard reagent in ether and hydrolysed by dilute acid, a carboxylic acid with one more carbon atom in the in is produced:

1: $C_2H_5MgI + CO_2 \rightarrow C_2H_5COOMgI$

2: $C_2H_5COOMgI + H^+ \rightarrow C_2H_5COOH + Mg^{2+} + I^-$

ethane gives ethyl magnesium iodide, which then produces propanoic acid.

Exam tip

If you cannot immediately see the route, try working backwards from the final product.

Knowledge check 19

State the reagents, conditions and intermediates formed in the preparation of N-phenylethanamide, $C_6H_5NHCOCH_3$ from benzene.

Methanal → primary alcohols

When methanal and a solution of a Grignard reagent in ether are mixed and then hydrolysed by dilute acid, a **primary alcohol** with a longer carbon chain is formed:

Step 1: $C_2H_5MgBr + HCHO \rightarrow C_2H_5CH_2OMgBr$

Step 2: $C_2H_5CH_2OMgBr + H^+(aq) \rightarrow C_2H_5CH_2OH + Mg^{2+} + Br^-$

Other aldehydes → secondary alcohols

When an aldehyde and a Grignard reagent, dissolved in ether, are mixed and then hydrolysed by dilute acid, a **secondary alcohol** with a longer carbon chain is formed:

Step 1: $C_2H_5MgI + CH_3CHO \rightarrow C_2H_5CH(OMgI)CH_3$

Step 2: $C_2H_5CH(OMgI)CH_3 + H^+ \rightarrow C_2H_5CH(OH)CH_3 + Mg^{2+} + I^-$

Ethylmagnesium iodide is converted to butan-2-ol. A carbon chain of two becomes a chain of four.

Ketones → tertiary alcohols

When a ketone and a Grignard reagent, dissolved in ether, are mixed and then hydrolysed by dilute acid, a **tertiary alcohol** is formed:

Step 1: $CH_3MgI + CH_3COCH_3 \rightarrow (CH_3)_3COMgI$

Step 2: $(CH_3)_3COMgI + H^+ \rightarrow (CH_3)_3COH + Mg^{2+} + I^-$

The product is the tertiary alcohol 2-methylpropan-2-ol.

Minimising risk in an experiment

Hazard data may be given in a question. This can be used to design an experiment that minimises the risks due to any hazards.

If a reactant or product:

- is poisonous or is a harmful or irritating gas, carry out the reaction in a fume cupboard
- is corrosive or is absorbed through the skin, wear disposable gloves
- is flammable, heat using a water-bath or an electric heater. This is essential if ether is present as the solvent of the reaction mixture, as in the reduction using lithium aluminium hydride.

Experimental techniques

You should be able to select suitable techniques for carrying out the reactions listed in Table 9 on page 50. You should also be able to draw the sets of apparatus shown on the following pages.

Heating under reflux

Heating under reflux is used when the reaction is slow and one of the reactants is volatile. The apparatus used is shown in Figure 52.

Exam tip

A hazard is a propert of a substance. The r depends on how that substance is handleo how the experiment i carried out.

Exam tip

The wearing of lab coats and eye protection is always assumed and will never score marks f minimising risk.

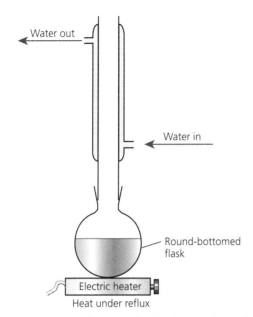

Figure 52 Apparatus used for heating under reflux

Exam tip

Always make sure that your diagrams of apparatus are not sealed, the water flow in a condenser is correct and that a round-bottomed (not flat) flask is used.

urification by washing

e product is distilled out of the reaction mixture and the following process is ried out.

The distillate is washed with sodium carbonate solution in a separating funnel. The pressure must be released from time to time to let out the carbon dioxide. This washing is repeated until no more gas is produced. This removes any acidic impurities that may be present.

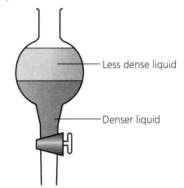

Figure 53 The layers formed by purification by washing

Exam tip

The denser liquid will form the lower layer. You may assume that a mixture of organic substances will form a single layer.

he aqueous layer is discarded and the organic layer washed with water. This emoves any unreacted sodium salts and any soluble organic substances, such as thanol.

he aqueous layer is discarded and the organic layer dried.

Solvent extraction

An organic product can often be separated from inorganic substances by solvent extraction. The compound 1-chloropentane can be prepared by adding excess phosphorus(v) chloride to pentan-1-ol.

$$CH_3(CH_2)_3CH_2OH(l) + PCl_5(s) \rightarrow CH_3(CH_2)_3CH_2Cl(l) + HCl(g) + POCl_3(l)$$

Phosphorus oxychloride ($POCl_3$) has a boiling temperature of 105°C, which is similar to that of 1-chloropentane at 108°C, so it would not be separated by distillation of the reaction mixture.

The organic compound is separated by adding ether (ethoxyethane). This solvent dissolves the 1-chloropentane but not the phosphorus oxychloride. The ether solution is washed to remove acid, dried and the ether removed by distillation.

Recrystallisation

At the end of a synthesis, the desired product may be impure because of the presence of unused reactants or the products of side reactions, so it needs to be purified.

The solid is filtered off from the reaction mixture and purified by recrystallisation.
- Select a solvent in which the solid is soluble when hot and almost insoluble at room temperature.
- Dissolve the solid in the minimum amount of hot solvent.
- Filter the solution through a pre-heated glass filter funnel fitted with a fluted filter paper — this removes any insoluble impurities.
- Allow the filtrate to cool and crystals of the pure solid to appear.
- Filter using a Buchner funnel under reduced pressure — this removes any soluble impurities.
- Wash the solid with a little cold solvent and allow to dry.

Exam tip

The details of recrystallisation and the type of impurities that are removed at each filtration are likely to be asked in paper

Drying

Organic liquids must be dried after separation. This is usually done with lumps of anhydrous calcium chloride. The cloudy liquid slowly goes clear as the water is absorbed by the drying agent.

Note that solid potassium hydroxide is used to dry amines and alcohols, as they form complex ions with calcium chloride.

Finally the organic liquid is decanted off from the drying agent and distilled, collecting the fraction that boils at ±2°C of the *Data Booklet* boiling temperature.

Distillation

This is used to remove a volatile substance from a mixture, e.g. a carboxylic acid from sulfuric acid and chromium compounds resulting from the oxidation of a primary alcohol. The apparatus used in distillation is shown in Figure 54.

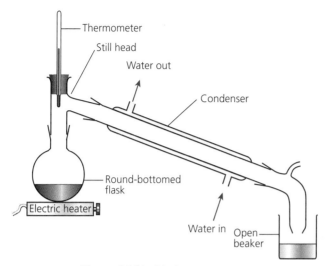

Figure 54 Distillation apparatus

eam distillation

am distillation is used to extract a volatile substance that is insoluble in water from action mixture that contains an immiscible liquid as well as a solution, for example nylamine after reduction of nitrobenzene with tin and concentrated hydrochloric l. It can also be used to extract volatile oils from natural products such as rose als or orange peel. The apparatus used in steam distillation is shown in Figure 55.

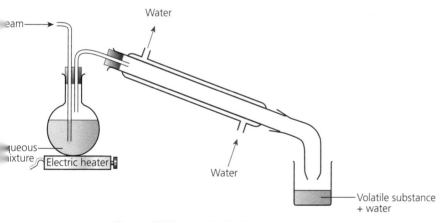

Figure 55 Steam distillation apparatus

In these diagrams it is always safer to draw an electric heater rather than a Bunsen burner, as a reactant or product might be flammable.

Make sure that each piece of the apparatus is drawn as a separate piece and not fused to the next one. The joints between the different parts are difficult to draw.

You must be able to draw a round-bottomed flask, a condenser (either in the reflux or distillation position) and a thermometer.

Make sure that the water goes *in* at the *bottom* of the condenser and *out* at the *top*.

Make sure that the apparatus is open to the air at some point — at the top of a reflux condenser or into a collecting vessel for distillation.

Melting and boiling-point determinations

Melting- and boiling-point determinations are used to identify a substance. The melting or boiling point is determined and checked against a data book.

Melting point

To determine the melting point heat the solid in a test tube immersed in a water-bath until the solid begins to melt. Measure the temperature at this point. When the substance has completely melted, allow it to cool, and then measure the temperature at which it begins to solidify. The substance must be stirred all the time. Calculate an average of the two temperatures.

Boiling point

To determine the boiling point of a substance use the following method:

- Place a small amount of the test liquid in an ignition tube and, using a rubber band, attach it to the thermometer.
- Place an empty capillary tube in the liquid, with its open end below the surface.
- Clamp the thermometer in the beaker of water.
- Slowly heat the water, stirring all the time. When the stream of bubbles coming out of the capillary tube is rapid and continuous, note the temperature of the water bath and stop the heating process.
- Allow the beaker of water to cool, stirring continuously. Note the temperature of the water bath when bubbles stop coming out of the capillary tube and the liquid begins to suck back into the capillary tube.

The average of these two temperatures is the boiling point of the liquid. The apparatus used is shown in Figure 56.

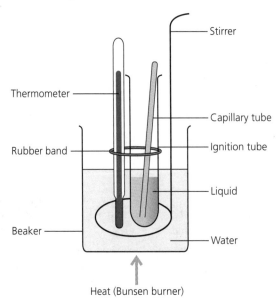

Figure 56 Apparatus used to determine boiling point

Exam tip

Make sure that you heat slowly in both melting and boiling point determinations not the reading will be too low.

Core practical 16

The preparation of aspirin

2-hydroxybenzoic acid can be converted into its ethanoate ester, which is shown in Figure 57.

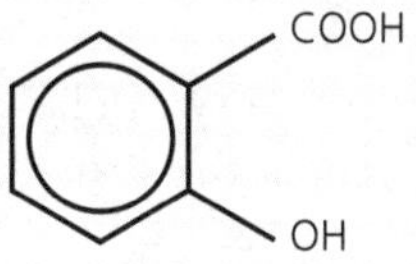

Figure 57 Ethanoate ester of 2-hydroxybenzoic acid

The ethanoate ester is produced by a reaction of 2-hydroxybenzoic acid with ethanoic anhydride, $(CH_3COO)_2O$. The product is aspirin. The reaction is shown in Figure 58.

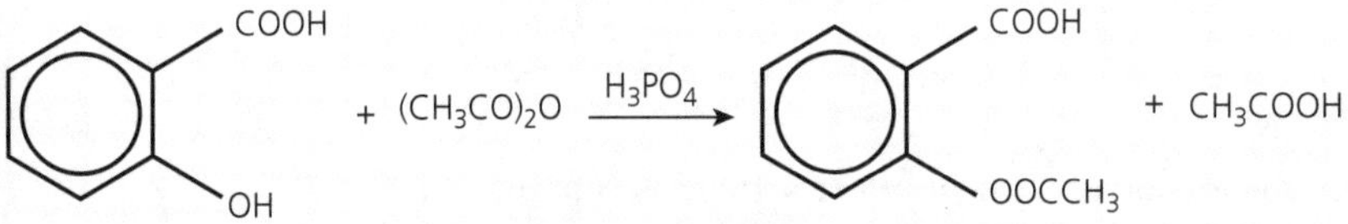

Figure 58 Reaction between 2-hydroxybenzoic acid and ethanoic anhydride

The method is as follows:

Place 5 g of 2-hydroxybenzoic acid and 10 cm^3 of ethanoic anhydride in a round-bottomed flask fitted with a reflux condenser.

Slowly add 3 cm^3 of concentrated phosphoric acid down the reflux condenser.

Heat the mixture using a water bath for 5 minutes and then add 5 cm^3 of water. This will hydrolyse any excess ethanoic anhydride.

Allow the mixture to cool and then pour it into a beaker containing about 100 cm^3 of cold water. If necessary scratch the inside of the beaker with a glass rod to start crystallisation and then cool the mixture in an ice bath.

Filter off the solid aspirin using a Buchner funnel and wash with a little cold water.

Finally recrystallise the aspirin using a minimum amount of boiling water.

Hazard: Ethanoic anhydride and concentrated phosphoric acid are corrosive, so gloves must be worn.

■ Topic 19 Modern analytic techniques II

Exam tip

Questions on Topic 7B infrared spectroscopy will be asked in A-leve papers 2 and 3.

Topic 19A Mass spectrometry

Topic 7A in the first year of the course described how mass spectrometry can be used to determine the molecular mass of a substances and how clues about the structure can be obtained from fragmentation patterns.

Worked example 1

An acidic organic compound X had a molecular ion peak at $m/z = 88$ and other major peaks at 73 and 45 but not one at 59 nor at 29. Identify the compound explaining your reasoning.

Answer

The peak at $m/z = 73$ is caused by the loss of a fragment of mass 15, so X contains one or more CH_3 groups.

The peak at $m/z = 45$ is due to $COOH^+$, so X is a carboxylic acid.

The lack of peaks at $m/z = 59$ and 29 shows that X does not contain a C_2H_5 group.

As the molar mass of $X = 88\,g\,mol^{-1}$, it is likely to be C_3H_7COOH. The lack of a C_2H_5 group shows that that is it $(CH_3)_2CHCOOH$, 2-methylpropanoic acid.

Use of four decimal place accuracy

Modern mass spectrometers can determine the m/z of the molecular ion to an accuracy of four decimal places. Atomic masses are also known to this accuracy and so the molecular formula of an unknown can be worked out. The substance is introduced into an accurate mass spectrometer and the m/z of the molecular ion measured. A computer then works out the molecular formula.

Worked example 2

A substance was thought to be either $HOOCCH_2COOH$ or $HOOC(CH_2)_2CH_2OH$. Its molecular ion had $m/z = 104.0471$.

Identify the substance. [Relative isotopic masses: $^{12}C = 12.0000$; $^1H = 1.0078$; $^{16}O = 15.9949$.]

Answer

$HOOCCH_2COOH$ would have a molecular ion of $m/z = 104.0108$

$HOOC(CH_2)_2CH_2OH$ would have a molecular ion of $m/z = 104.0471$, so the substance is $HOOC(CH_2)_2CH_2OH$, 4-hydroxybutanoic acid

opic 19B Nuclear magnetic resonance

he nuclei of all atoms are spinning. If the atom has an odd number of protons
 an odd number of neutrons (but not both) the atom will behave as a magnetic
pole. The magnetic field of this dipole can be aligned either parallel to or opposed
 an applied magnetic field. ^1H and ^{13}C nuclei produce a weak magnetic field.
hen they are placed in a strong magnetic field, the nucleus's own magnetic field
n lie parallel with the applied magnetic field or opposed to it. There is a slight
ergy difference between these two states. The nuclei can then absorb radio waves
d move from the lower (parallel) state to the higher energy (antiparallel) state.
is is shown in Figure 59.

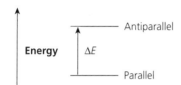

Figure 59 The energy difference between the parallel and antiparallel states

e frequency of radio waves absorbed depends upon the chemical environment of
 ^1H atoms (protons) or the ^{13}C atoms in the molecule. The amount that this differs
m a reference substance is called the **chemical shift**, δ.

C NMR spectroscopy

tural carbon contains 1.1% of the carbon-13 isotope. This, unlike the carbon-12
tope, has a magnetic dipole and so will absorb radio waves when the two levels are
t by a strong external magnetic field. In an organic compound the electron cloud
und a particular carbon atom will be slightly distorted due to the difference in the
tial charge on the carbon atom caused by what it is bonded to. Thus the chemical
ft varies from carbon atom to carbon atom. The *Data Booklet* has a list of chemical
ts and this will be available in exams.

tan-2-one, $CH_3COCH_2CH_2CH_3$, will have five peaks due to each of the five
bon atoms, as each has a slightly different chemical environment. However,
tan-3-one, $CH_3CH_2COCH_2CH_3$ will only have three peaks as both CH_3
ups are in the same chemical environment and both CH_2 groups are also in
al chemical environments. The ability to calculate the number of carbon atoms
nique environments is vital. 2-methylpropanoic acid has carbon atoms in three
que environments and so has three peaks in its ^{13}C NMR spectrum. Butanoic acid
 carbon atoms in four unique environments and so has four peaks in its ^{13}C NMR
ctrum.

Worked example 1

The alcohol C_3H_7OH exists as two isomers. One of these had the ^{13}C NMR spectrum shown below. Identify the isomer in Figure 60.

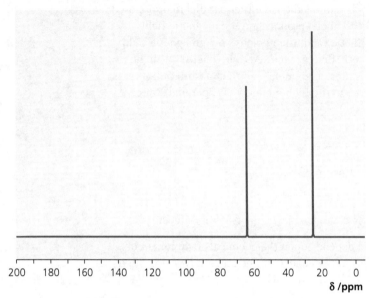

Figure 60 ^{13}C NMR spectrum of the alcohol C_3H_7OH

Exam tip

The peak at 25 ppm is due to the CH_3 carbon atoms and the peak at 64 ppm is due to the carbon of the C-OH.

Answer

As there are three carbon atoms but only two peaks, two of the carbon atoms must be in the same chemical environment. The compound is propan-2-ol, $CH_3CH(OH)CH_3$.

^1H NMR (proton NMR) spectroscopy

The theory underlying proton NMR spectroscopy is the same as that for ^{13}C NMR. The first question that has to be answered is to work out the number of unique environments there are for the hydrogen atoms in the molecule. This will equal the number of peaks in the spectrum.

Thus propanone (CH_3COCH_3) has a single peak, as the chemical environments of the hydrogen nuclei in the two CH_3 groups are identical, whereas propanal (CH_3CH_2CHO) has three peaks, one due to the CH_3 hydrogens, one due to the CH_2 hydrogens and one due to the CHO hydrogen. These peaks have heights in the ratio of 3:2:1, as there are three H nuclei in the CH_3 group, two in the CH_2 group and just one in the CHO group. The value of the chemical shift, δ, depends on the extent that the electrons around the hydrogen atom are drawn away by neighbouring groups. This is called deshielding. A neighbouring C=O group causes considerable deshielding and so a large value of δ. Chemical shift values are to be found in the *Data Booklet*. That for the hydrogen of a CHO group has a shift of between 9 and 10 relative to the standard.

Exam tip

You must be able to use the *Data Booklet* to identify hydrogen environments by their δ values.

he spectra produced by modern machines are more complicated. The magnetic
elds of hydrogen atoms on neighbouring *carbon* atoms slightly alter the applied
agnetic field felt by the hydrogen nucleus. This is called spin coupling and causes a
litting of the peak according to the $(n + 1)$ rule.

If there is *one* hydrogen atom on a neighbouring carbon atom, the peak is split into
$(1 + 1) = two$ peaks.

If there are *two* hydrogen atoms on neighbouring carbon atoms, the peak is split
into *three* and, if there are three neighbouring hydrogen atoms, the peak is split into
four, etc.

If there are no neighbouring hydrogen atoms, the peak is not split. Thus the peak
due to the CH_3 groups in propanone (CH_3COCH_3) is not split.

The peak due to the hydrogen atom of OH or NH groups is never split and does
not cause splitting of other hydrogen nuclei, so the unsplit peak in the proton NMR
spectrum of ethanol is due to the OH hydrogen nucleus.

us, as you can see in Figures 61 and 62:

The peak due to the CH_3 hydrogen nuclei in propan-2-ol, $CH_3CH(OH)CH_3$, is
split into two by the CH hydrogen.

The peak due to the hydrogen nucleus in the CH group in propan-2-ol is split into
seven as it has six hydrogen atoms on neighbouring carbon atoms (three in each
CH_3 group).

The peak due to the CH_3 hydrogen nuclei in ethanol, CH_3CH_2OH, is split into
three by the two hydrogen nuclei in the CH_2 group and the peak due to the CH_2
hydrogen nuclei is split into four.

Exam tip

The ^1H NMR spectrum
must be explained in
terms of the number
of different hydrogen
environments, the
peak heights and δ
values, and the splitting
pattern.

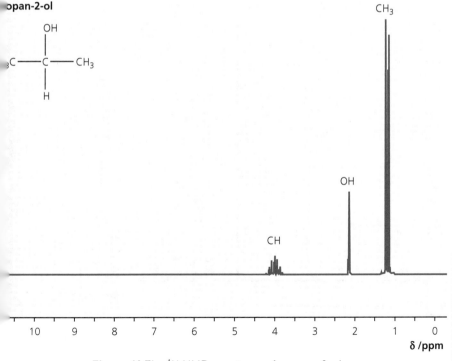

Figure 61 The ^1H NMR spectrum of propan-2-ol

The images below show the ¹H NMR spectra for ethanol (Figure 62) and propanone (Figure 63).

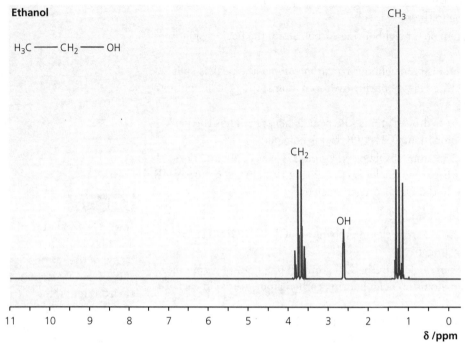

Figure 62 The ¹H NMR spectrum of ethanol

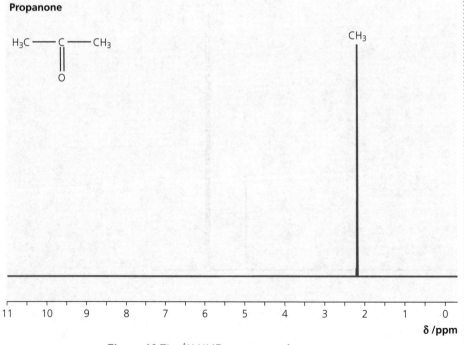

Figure 63 The ¹H NMR spectrum of propanone

he same technique is used to study the different environments of water in soft
man tissue, such as the brain, spinal discs or cartilage in joints. This is called
agnetic resonance imaging or MRI.

tegrated peak heights

me spectra show the relative areas under each set of peaks using another line
led the integrator trace. The area under this line is proportional to the number of
drogen atoms in that environment. Thus the relative heights of the integrator trace
the three hydrogen environments in propanoic acid are in the ratio 1:2:3. This is
own in blue in Figure 64.

Knowledge check 21

Use the *Data Booklet* to
find the δ values of the
peaks in the ^{1}H NMR
spectrum of propanal
and suggest the
splitting pattern of the
three peaks.

Knowledge check 22

Explain why there is no
splitting of the peaks in
^{13}C NMR spectroscopy.

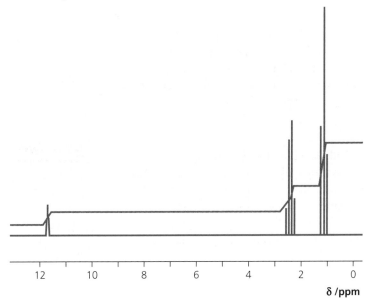

Figure 64 ^{1}H NMR spectrum from propanoic acid showing the integrator trace

e that the height of the peaks in a ^{13}C NMR spectrum has nothing to do with the
mber of carbon atoms in that environment.

pic 19C Chromatography

forms of chromatography have a stationary phase. This can be a solid packed into
lumn, a thin layer of solid on a support or a liquid adsorbed on a solid matrix.

mixture then passes through the stationary phase with an eluent, which is either
quid or a gas. The different components travel through the stationary phase at
erent speeds and so are separated.

speed at which a specific substance passes through depends on the nature of
stationary phase and of the eluent. For a given stationary phase and eluent, this is
sured by the R_f value of the substance:

$$R_f \text{ value} = \frac{\text{distance moved by substance}}{\text{distance moved by eluent}}$$

Exam tip

Different substances
have different
strengths of
intermolecular forces
between them and
the stationary and
moving phases
and so pass through
at different speeds.

High-performance liquid chromatography

High-performance liquid chromatography (HPLC) consists of a column packed with a solid of uniform particle size — the stationary phase. The sample to be separated is dissolved in a suitable solvent and added at the top of the column. The liquid eluent — the moving phase — is then forced through the column under high pressure.

The time taken for a component in the sample mixture to pass through the column is called the retention time and is a unique characteristic of the substance, the composition of the eluent, the nature of the stationary phase and the pressure. This means that different components will pass through one after the other with gaps between each.

The use of high pressure increases the speed at which the eluent passes through the column and so reduces the extent to which the band of a component spreads out due to diffusion. This gives it a much higher resolution than paper or thin-layer chromatography. The column can be connected to an infrared or mass spectrometer, which can then identify each component in the mixture.

Gas–liquid chromatography

Gas–liquid chromatography involves a sample of a liquid mixture being injected into a chromatographic column that is in a thermostatically controlled oven. The sample evaporates and is forced through the column by the flow of nitrogen gas or another inert, gaseous eluent. The column itself contains a liquid stationary phase that is adsorbed onto the surface of an inert solid. The gases coming out pass through a detector — usually a device that measures the thermal conductivity of the gas.

The identity of the components in the mixture can be found by removing samples leaving the column and measuring their infrared or mass spectra. These are then compared with spectra in a database.

Thin-layer chromatography

This is used to separate amino acids and the technique is described on page 46.

Infrared spectroscopy

This was dealt with in topic 7B in the second student guide of this series. Questions in papers 2 and 3 of A-level will be asked about a mixture of data from a mass spectrometer and an infrared spectrum.

Exam tip

The rate at which a substance passes through the column depends on the strength of the intermolecular forces between it and the liquid stationary phase

---

Knowledge check

The mass and infrared spectra of a compound $C_4H_8O_2$ were measured. The results were:

Mass spectrum: m/z of molecular ion = 88, other peaks at 57 and 43

IR spectrum: peaks at $3410\,cm^{-1}$ and at $1710\,cm^{-1}$

Suggest a skeletal formula for the compound. Justify your answer.

Summary

After studying this topic, you should be able to:
- use the *Data Booklet* to interpret mass, NMR and infrared spectra
- explain why hydrogen nuclei absorb radio frequencies when in a strong magnetic field
- relate the number of NMR peaks, their δ values, their peak heights and their splitting patterns to a suggested structure
- relate the frequencies of peaks in an infrared spectrum to a suggested structure
- identify the fragments causing peaks in a mass spectrum
- explain how GC and HPLC work and for what types of mixture each technique is used

Questions & Answers

is section contains multiple-choice, short open, calculation and extended writing
estions similar to those you can expect to find in papers 2 and 3. The questions
en here are not balanced in terms of types of questions or level of demand. In the
m, the answers are written on the question paper. Here, the questions are not
wn in exam paper format.

e answers given are those that examiners would expect from a grade-A candidate.
ey are not 'model answers' to be regurgitated without understanding. In answers
: require extended writing, it is usually the ideas that count rather than the form of
ds used. The principle is that correct and relevant chemistry scores.

nments on the questions are indicated by the icon ℮. They offer tips on what you
d to do to get full marks. Comments on the student answers are preceded by the
℮. These comments may explain the correct answer, point out common errors
e by students who produce work of C-grade standard or lower, or contain other
ul advice.

e exam papers

questions on paper 2 are based mainly on kinetics and organic chemistry. There
also be questions on bonding and structure (topic 2) and formulae, equations and
unts of substance (topic 5) covered in the first student guide of this series and
nic chemistry I (topic 6) and modern analytical techniques I (topic 7) covered in
econd student guide of this series.

r 2 allows 1 hour 45 minutes and has a maximum of 90 marks.

r 3 is fully synoptic with questions on all the topics in the specification. It lasts
urs 30 minutes and has a maximum of 120 marks. It will contain one question
d on a piece of stimulus material, which could be up to one page. A significant
ortion of the questions will be based on practical experiments that should have
covered in both years of the course, and especially on the core practicals.

nmand terms

niners use certain words that require you to respond in a specific way. You must
guish between these terms and understand exactly what each requires you to do.

efine — give a simple definition without any explanation.

entify — give the name or formula of the substance.

ate — no explanation is required (and you shouldn't give one).

educe — use the information supplied in the question to work out your answer.

uggest — use your knowledge and understanding of similar substances, or those
th the same functional groups, to work out the answer.

ompare — make a statement about *both* substances being compared.

- **Explain** — use chemical theories or principles to say why a particular property is as it is.
- **Predict** — say what you think will happen on the basis of the principles that you have learned.
- **Justify your answer** — give a brief explanation of why your answer is correct.

Organic formulae

- **Structural formula** — you must give an unambiguous structure. For instance, $CH_3CH_2CH_2OH$ or $C_2H_5CH_2OH$ are acceptable, but C_3H_7OH could be either propan-1-ol or propan-2-ol and so is not acceptable. If a compound has a double bond, then it is better to show it in a structural formula.
- **Displayed or full structural formula** — you must show all the *atoms* and *all* the *bonds*. 'Sticks' instead of hydrogen atoms will lose marks.
- **Skeletal formula** — remember that there is a carbon atom at the end of each line and at 'joints' in the skeleton.
- **Shape** — if the molecule or ion is pyramidal, tetrahedral or octahedral you must make sure that your diagram looks three-dimensional by using wedges and dashes. Draw optical isomers as mirror images of each other. Geometric isomers must be drawn with bond angles of 120°. Make sure that the *bonds go to the correct atom* — for example, the oxygen in an –OH group or the carbon in $-CH_3$ and –COOH groups.

Points to watch

- **Stable** — if you use this word, you must qualify it — for example, 'stable *to heat*', 'the reaction is *thermodynamically* stable', 'the reaction is *kinetically* stable' or 'a secondary carbocation intermediate is *more* stable *than* a primary carbocation'.
- **Reagents** — if you are asked to identify a reagent, you must give its *full* name or formula. Phrases such as 'acidified dichromate(vi)' will not score full marks. You must give the reagent's full name — for example, 'potassium dichromate(vi)'.
- **Conditions** — don't use abbreviations such as 'hur' for heat under reflux.
- **Atoms, molecules and ions** — don't use these words randomly. Ionic compounds contain ions, not molecules.
- **Rules** — don't use rules such as Markovnikov or Le Châtelier to *explain*. However, they can be used to predict.

Approaching the A-level papers

Revision

- Start your revision in plenty of time. Make a list of what you need to do, emphasising the topics that you find most difficult, and draw up a detailed revision plan. Work back from the examination date, ideally leaving an entire week free from fresh revision before that date. Be realistic in your revision plan and then add 25% to the timings because everything takes longer than you think.
- When revising, make a note of difficulties and ask your teacher about them. If you do not make these notes, then you may forget to ask.

Make use of past papers. Similar questions are regularly asked, so if you work through as many past papers and answers as possible, then you will be in a strong position to obtain a top grade.

When you use the Questions & Answers section of this guide, make a determined effort to write your answers *before* looking at the sample answers and examiner comments.

e exam

Read the question. Questions change from one examination to the next. A question that looks the same, at a cursory glance, to one that you have seen before usually has significant differences when read carefully. Needless to say, students do not receive credit for writing answers to their own questions.

Be aware of the number of marks available for a question. This is an excellent pointer to the number of things you need to say.

Do not repeat the question in your answer. The danger is that you fill up the space available and think that you have answered the question, when in reality some or maybe all of the real points have been ignored.

Look for words in **bold** in a question and make sure that you have answered the question fully in terms of those words or phrases. For example, if the question asks you to define the **standard electrode potential**, make sure that you explain the meaning of electrode potential as well as what the standard conditions are.

Questions at A-level may involve substances or situations that are new to you. This is deliberate and is what makes these questions AO3. Don't be put off by large organic molecules. They are nothing more than a collection of functional groups that you can assume react independently of each other.

following list of potential pitfalls to avoid is particularly important:

Do not write in any colour other than black. This is an exam board regulation. The scans are entirely black-and-white, so any colour used simply comes out black — unless you write in red, in which case it does not come out at all. The scanner cannot see red (or pink or orange) writing. So if, for example, you want to highlight different areas under a graph, or distinguish lines on a graph, you must use a different sort of shading rather than a different colour.

Do not use small writing. Because the answer appears on a screen, the definition is slightly degraded. In particular, small numbers used for powers of 10 can be difficult to see. The original script is always available but it can take a long time to get hold of it.

Do not write in pencil. Faint writing does not scan well.

Do not write outside the space provided without saying, within that space, where the remainder of the answer can be found. Examiners only have access to a given item; they cannot see any other part of your script. So if you carry on your answer elsewhere but do not tell the examiner within the clip that it exists, it will not be seen. Although the examiner cannot mark the out-of-clip work, the paper will be referred to the Principal Examiner responsible for the paper.

Do not write across the centre-fold of the paper from the left-hand to the right-hand page. A strip about 8 mm wide is lost when the papers are guillotined for scanning.

■ Paper 2 and paper 3-type questions

Each multiple-choice question, or incomplete statement, is followed by four suggested answers, A, B, C or D. Select the *best* answer in each case.

For other questions it is worth reading through the whole question before attempting to answer it.

Question 1

(a) What is meant by the term **order of reaction**? (1 mark)

(b) For the reaction between compounds X and Y, the rate equation is:

   rate = $k[X]^2[Y]$

   The units of rate are $mol\,dm^{-3}\,s^{-1}$. The units of the rate constant, k, for this reaction are:

   A $mol^{-1}\,dm^3\,s^{-1}$

   B $mol\,dm^{-3}\,s$

   C $mol^{-2}\,dm^6\,s^{-1}$

   D $mol^2\,dm^{-6}\,s$ (1 mark)

(c) 1-bromopropane and sodium hydroxide react to form propan-1-ol and sodium bromide. The rate of this reaction was measured using different concentrations of hydroxide ions and 1-bromopropane. The results are shown in Table 1.

Table 1 The rates of reaction using different concentrations of hydroxide ions and 1-bromopropane

Experiment	$[OH^-]/mol\,dm^{-3}$	$[1\text{-bromopropane}]/mol\,dm^{-3}$	Initial rate of reaction/$mol\,dm^{-3}\,s^{-1}$
1	0.25	0.25	1.4×10^{-4}
2	0.125	0.25	7.0×10^{-5}
3	0.50	0.125	1.4×10^{-4}

   (i) Deduce the order of reaction with respect to hydroxide ions and to 1-bromopropane. Justify your answer. (4 marks)

   ⊜ Make sure that you identify clearly the experiments you are considering and whether the concentration of one or both substances changes. Beware when finding the order for 1-bromopropane because both concentrations alter between experiments 2 and 3.

   (ii) Write the rate equation for this reaction. (1 mark)

   (iii) Calculate the value of the rate constant. (1 mark)

   (iv) Write a mechanism for this reaction that is consistent with your answer to (ii). (3 marks)

   ⊜ A curly arrow must start on a bond or on a lone pair of electrons. It must go either to an atom to form an ion or towards an atom to form a bond. Make sure that you do not omit any charges.

(Total: 11 marks)

tudent answers

) The sum of the powers to which the *concentrations* of the reactants are raised in the experimentally determined rate equation ✓.

Do not say that it is the sum of the individual orders, unless you also define ividual order.

) C ✓

The dimensions of k are $\dfrac{\text{mol dm}^{-3}\,\text{s}^{-1}}{(\text{mol dm}^{-3})^3} = \text{mol}^{-2}\,\text{dm}^6\,\text{s}^{-1}$

) (i) Consider experiments 1 and 2. When [OH⁻] is halved and [1-bromopropane] is kept constant, the rate also halves. Therefore, the order with respect to hydroxide ions is 1 ✓.

Consider experiments 1 and 3. When [OH⁻] is doubled and [1-bromopropane] is halved, the rate does not alter. Since the rate should double because of the doubling of [OH⁻] ✓, the halving of [1-bromopropane] must have caused the rate to halve ✓, cancelling out the increase due to the rise in [OH⁻]. Therefore, the reaction is first order ✓ with respect to 1-bromopropane.

You must make clear which two experiments you are comparing.

(ii) rate = k[OH⁻][1-bromopropane] ✓

(iii) k = rate/[OH⁻][1-bromopropane] = $\dfrac{1.4 \times 10^{-4}}{0.25 \times 0.25}$ = 0.0022 ✓ mol⁻¹ dm³ s⁻¹

(iv) The rate equation is first order with respect to both reactants, so the mechanism is S_N2. It is attacked by the nucleophilic OH⁻ ion and involves a transition state.

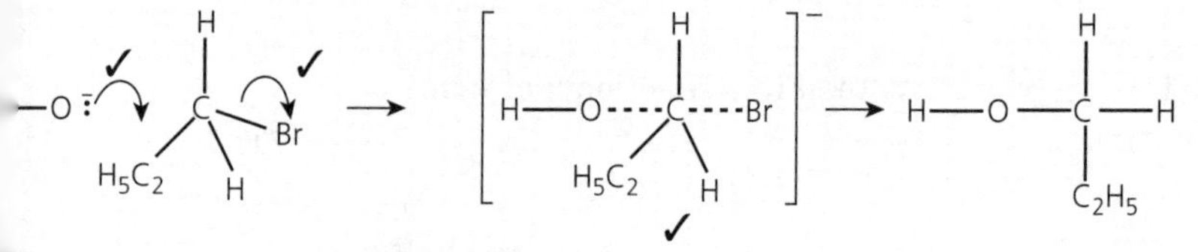

he curly arrow should start from the lone pair of electrons on the oxygen ı of the OH⁻ ion and go towards the δ+ carbon atom. A second curly arrow ᵗ be drawn from the σ C–Br bond to the δ– bromine atom. The transition state ᵗ show a negative charge.

Question 2

(a) The rate of a reaction is best expressed as:

    A   the time taken for about 10% of the reaction to take place

    B   the time taken for the reaction to finish

    C   1/time for about 10% of the reaction to take place

    D   1/the time for the reaction to finish   (1 mark)

(b) When dilute acid is added to a solution of $S_2O_3^{2-}$ ions, a precipitate of sulfur slowly forms:

$$S_2O_3^{2-}(aq) + 2H^+(aq) \rightarrow S(s) + SO_2(aq) + H_2O(l)$$

The rate of this reaction can be studied by recording the time taken for the amount of sulfur to be produced which hides a cross marked underneath a beaker. If the experiment is repeated at different temperatures, the activation energy can be calculated.

(i) The rate constant, k, is related to the temperature by:

$\ln k = \dfrac{-E_a}{RT} +$ a constant

and k is itself proportional to 1/time.

Complete Table 2 and use the data to plot a graph of $\ln(1/\text{time})$ against 1/temperature. Measure the gradient of the line and hence calculate the activation energy for this reaction. Include units with your answer. [$R = 8.31\,\text{J}\,\text{K}^{-1}\,\text{mol}^{-1}$]   (7 marks)

Table 2

Temperature/K	1/T/K⁻¹	Time/s	1/time/s⁻¹	ln 1/time
288		53		
298		32		
308		20		
318		13		

(e) Use a sensible scale for the graph. There is no point in having the origin on your axes. Both axes should start from just less than the smallest value in the table.

(ii) Explain, in terms of kinetic theory, why the rate of this reaction increases as the temperature is increased.   (3 marks)

Total: 11 marks

Student answers

(a) C ✓

(b) (i)

Temperature/K	1/T/K⁻¹	Time/s	1/time/s⁻¹	ln 1/time
288	0.00347	53	0.0189	−3.97
298	0.00336	32	0.03125	−3.45
308	0.00325	20	0.0500	−3.00
318	0.00314	13	0.0769	−2.53
	all correct ✓			all correct ✓

Axes labelled ranging from −4 to −2.5 and 0.003(1) to 0.0035 ✓

Points plotted ✓

slope (gradient) = −1.44/0.00033 = −4364 ✓

E_a = −R × slope = −8.31 × (−4364) = +36 261 J mol⁻¹ = + 36 kJ mol⁻¹ value ✓ unit ✓

) An increase in temperature causes the molecules to gain kinetic energy ✓. As a result, more of the colliding molecules have energy ≥ the activation energy ✓, so a greater *proportion* of the collisions result in reaction ✓ (i.e. more of the collisions are successful). Hence the rate of reaction increases.

For the third mark, do not say that 'There will be more successful collisions'. ₂ total number of successful collisions is independent of the temperature ause the reaction will still go to completion, but over a longer time, at a lower nperature.

ıestion 3

Under suitable conditions, both propanoic acid and propanoyl chloride react with methanol to form an ester.

(i)   Name the ester produced.   (1 mark)

(ii) Write equations for the formation of this ester from:
   ■ propanoic acid
   ■ propanoyl chloride   (2 marks)

Make sure that you have the correct arrow for each equation. Is it → or ⇌?

(iii) Which method gives a better yield of ester — the reaction with propanoyl chloride or the reaction with propanoic acid? Explain your answer.   (1 mark)

(iv) The correct repeat unit for the polyester formed by reacting hexane-1,6-dioic acid with hexane-1,6-diol is:

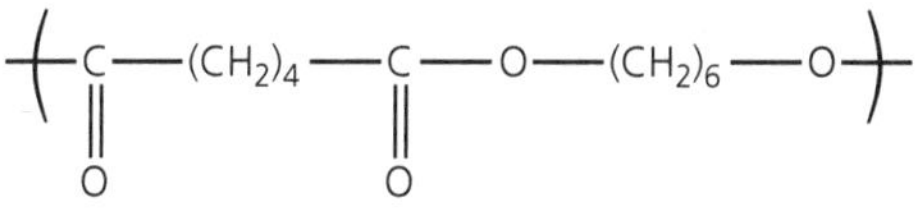

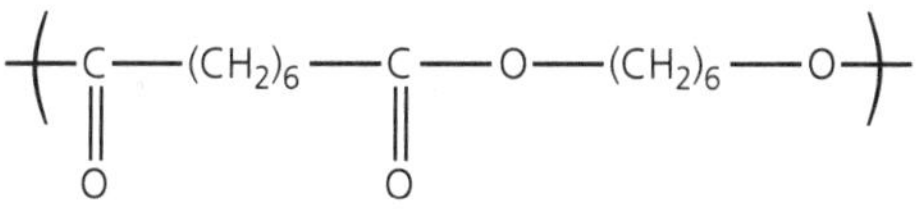

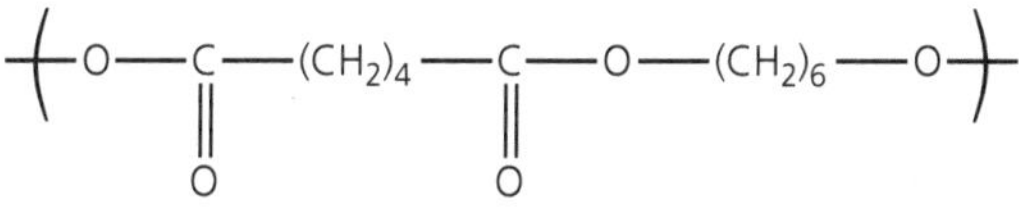

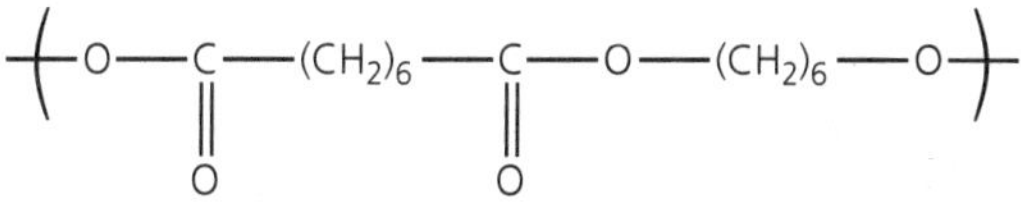

                                                                                                    (1 mark)

(v) Poly(ethenol) has the repeat unit $CH_2-CH(OH)$. It is:

    A made by polymerising ethenol at room temperature

    B made by polymerising ethenol using a Ziegler–Natta catalyst

    C made by hydrolysis of polyvinyl acetate with dilute acid

    D insoluble in water because it cannot form enough hydrogen bonds with water molecules **(1 mark)**

(b) Vegetable oils can be converted into biodiesel by a transesterification reaction. Give one example of this. **(2 marks)**

e You should give an equation, either in symbols or in words, and say which of the products constitute biodiesel.

(c) Ethane-1,2-diol and benzene-1,4-dicarboxylic acid form a polyester. Write the structural formula of the repeat unit of this polymer showing all double bonds. **(2 marks)**

e Check the number of oxygen atoms that you have written in the repeat unit.

Total: 10 marks

Student answers

(a) (i) Methyl propanoate ✓

    (ii) With propanoic acid:

        $CH_3CH_2COOH + CH_3OH \rightleftharpoons CH_3CH_2COOCH_3 + H_2O$ ✓

    With propanoyl chloride:

        $CH_3CH_2COCl + CH_3OH \rightarrow CH_3CH_2COOCH_3 + HCl$ ✓

    (iii) The reaction with propanoyl chloride because it is not a reversible reaction whereas the reaction with propanoic acid is reversible ✓.

    (iv) A ✓

e Hexane-1,6-dioic acid has six carbon atoms (including the carboxylic acid carbon atoms), so options B and D are incorrect as they both have eight carbon atoms in the dioic acid parts of their chains. The polyester forms by loss of an OH from one monomer and an H from the other, so the repeat unit has only four oxygen atoms. This means that options C and D are incorrect because they have five oxygen atoms in their repeat units.

    (v) C ✓

e Ethenol ($CH_2=CHOH$) does not exist. Attempts to make it produce ethanal (CH_3CHO), so options A and B are incorrect. Poly(ethenol) is water soluble, so option D is wrong. It is manufactured either by transesterification with methanol or by the hydrolysis of polyvinyl acetate ($(H_2CCH(OOCCH_3))_n$) so C is the correct answer.

b) The transesterification reaction involved is when the ester of propane-1,2,3-triol and unsaturated long-chain fatty acids reacts with an alcohol such as methanol. This produces three methyl esters of the unsaturated acids, which are used as biodiesel ✓, and a by-product of propane-1,2,3-triol:

$$CH_3OOCR$$
$$|$$
$$CH_3OOCR' + 3CH_3OH \longrightarrow RCOOCH_3 + R'COOCH_3 + R''COOCH_3 + CHOH \checkmark$$
$$|$$
$$CH_3OOCR''$$

with

$$CH_2OH$$
$$|$$
$$CHOH$$
$$|$$
$$CH_2OH$$

c)

Ester linkage ✓; rest of repeat unit with 'continuation' bonds ✓. The repeat unit a polyester has four oxygen atoms.

Question 4

This question is about carbonyl compounds.

An isomer of C_5H_8O gave a yellow precipitate with iodine and alkali, decolorised bromine water and gave a yellow precipitate with 2,4-dinitrophenylhydrazine. The isomer is:

A $CH_2=C=CHCH(OH)CH_3$
B $CH_3CH=CHCH_2CHO$
C $CH_2=CHCH_2COCH_3$
D $CH_2=CHCOCH_2CH_3$ (1 mark)

Propanone and propanal are isomers with the molecular formula C_3H_6O.

(i) In which of the following is there hydrogen bonding?
    A   between different molecules of propanone
    B   between different molecules of propanal
    C   between propanal and propanone molecules
    D   between propanal molecules and water (1 mark)

(ii) State the observation of the reaction of propanone with 2,4-dinitrophenylhydrazine. (1 mark)

(iii) Write the formula of the organic product of the reaction of propanal with Fehling's solution and that with lithium aluminium hydride in dry ether followed by the addition of dilute hydrochloric acid. (2 marks)

Is Fehling's solution acidic, alkaline or neutral?

(i) Both propanone and propanal react with hydrogen cyanide, HCN, in the presence of a catalyst of cyanide ions. In the nucleophilic addition of HCN to propanone, the first step is attack by a lone pair of electrons:

Questions & Answers

    A   on the oxygen atom of propanone onto the δ+ hydrogen atom of HCN

    B   on the carbon atom of HCN onto the δ+ carbon atom in propanone

    C   on the carbon atom of the CN⁻ ion onto the δ+ carbon atom in propanone

    D   on the nitrogen atom of the CN⁻ ion onto the δ+ carbon atom in propanone (1 mark)

(ii) Draw the mechanism of the reaction with propanone.   (3 marks)

ℯ Your mechanism should be written in two steps. CN⁻ ions are a catalyst, so they should be on the left-hand side of step 1 and re-formed in step 2. As always, include signs on ions.

(iii) Explain why this reaction is very slow at both high and low pH.   (2 marks)

ℯ HCN is a weak acid, so CN⁻ ions are a base. Use this and your mechanism to answer the question, remembering that the rate of a reaction depends on the concentration of the reactants.   Total: 11 marks

Student answers

(a) C ✓

ℯ The iodoform reaction tests for CH₃C=O or a CH₃CH(OH) group. Bromine water tests for a C=C group and Brady's reagent for a C=O group. The question is best attempted by putting a ✓ or ✗ for each reaction.

	CHI₃ test	Br₂(aq)	2,4-DNP
A	✓	✓	✗
B	✗	✓	✓
C	✓	✓	✓
D	✗	✓	✓

(b) (i) D ✓

ℯ For hydrogen bonding to occur there must be a δ+ hydrogen atom in one molecule and a δ− oxygen atom in the other molecule. Propanal has a δ− oxygen (caused by the considerable difference in electronegativities of the carbon and the oxygen in the C=O group) and there are δ+ hydrogen atoms in the water. This means that hydrogen bonding will occur. Neither propanone nor propanal has a hydrogen atom that is sufficiently δ+ for hydrogen bonding, so there are no hydrogen bonds between propanone molecules (option A), between propanal molecules (option B) or between propanal and propanone molecules (option C).

(ii) A yellow/orange precipitate ✓

(iii) CH₃CH₂COO⁻ ✓ and CH₃CH₂CH₂OH ✓

ℯ Fehling's solution is alkaline and oxidises propanal to give the anion of propanoic acid, but not the acid itself.

(c) (i) C ✓

This reaction requires a catalyst of cyanide ions. The formula of a cyanide ion usually written as CN^-. However, the charge is on the carbon atom, which has a ne pair of electrons. In the first step, the lone pair attacks the δ+ carbon atom in ropanone, forming a new carbon–carbon bond.

(ii)

Step 1

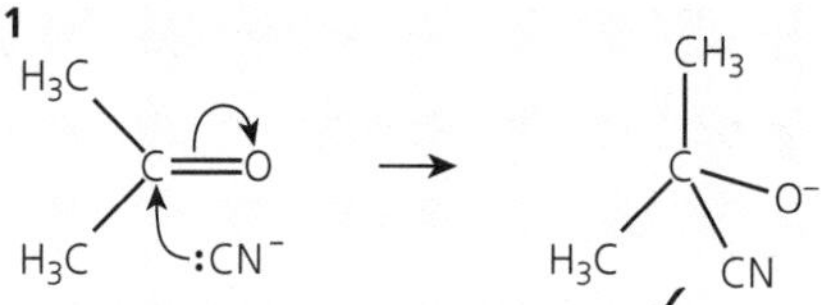

both arrows ✓

Step 2

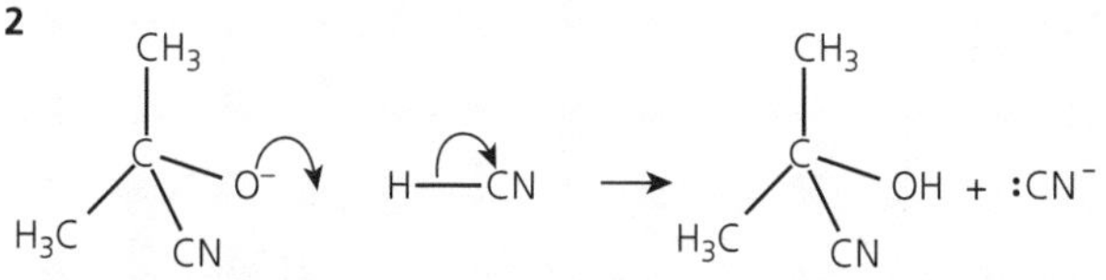

both arrows ✓

(iii) At a high pH (very alkaline), the OH^- ions react with the HCN. As the [HCN] is now very small, the rate of step 2 is almost zero ✓. At a low pH (very acid), the H^+ ions protonate any CN^- ions. As the $[CN^-]$ is now very small, the rate of step 1 is almost zero ✓.

uestion 5

enes, such as benzene and its derivatives nitrobenzene and phenol, react with ctrophiles.

Nitrobenzene can be prepared from benzene. Give the reagents and conditions and write the equation for this reaction. Why must the temperature be controlled?　　　　　　　　　　　　　　　　　　(5 marks)

Give the names or formulae of the reagents and whether they are aqueous concentrated and the temperature needed. State what would happen if the nperature rose or fell too much.

Benzene also reacts with ethanoyl chloride in a Friedel–Crafts reaction. Give the mechanism for this reaction, including the production of the electrophile.　(4 marks)

Make sure the charge on the electrophile is on the correct atom and that curly arrows start and finish in the correct places. Have you drawn the rmediate carefully?

Phenol reacts with bromine water.

(i) Write the equation for this reaction.　　　　　　　　　　　　　(1 mark)

Remember that this is a substitution reaction and so there will be two ducts.

(ii) Explain why this reaction does not need a catalyst and is much faster than the reaction of bromine with benzene

(3 marks)

ⓔ You must comment on the effect of the OH group on the delocalised electrons in the ring and how this makes this type of reaction faster.

Total: 13 marks

Student answers

(a) The reagents are concentrated ✓ nitric and sulfuric acids ✓.

The conditions are heat to 50°C ✓. At a higher temperature some dinitrobenzene is produced and at a lower temperature the reaction is too slow ✓. The equation is:

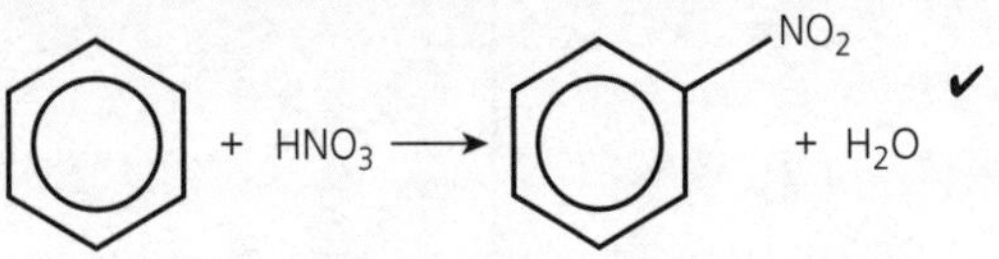

(b) Formation of electrophile:

$$CH_3COCl + AlCl_3 \rightarrow CH_3C^+{=}O + AlCl_4^-$$

Step 1

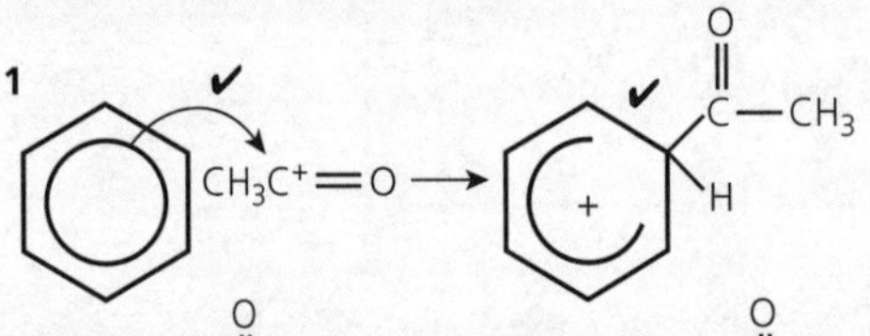

Step 2

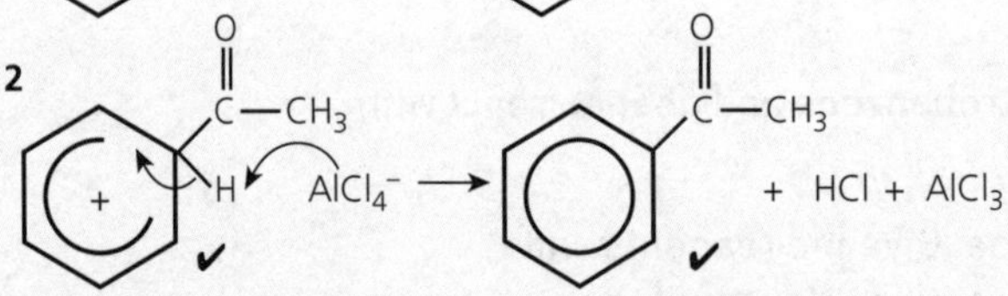

ⓔ Make sure that the delocalised ring is broken in the intermediate and that it is +. The mark for step 2 would be awarded without the $AlCl_4^-$ being shown and with H^+, rather than $HCl + AlCl_3$, as a product.

(c) (i)

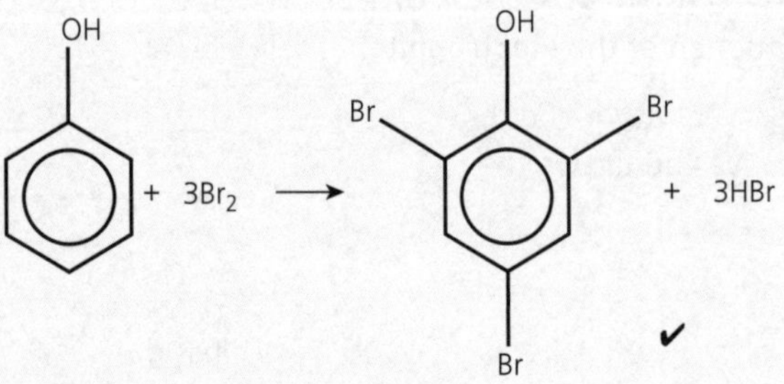

(ii) The lone pair of electrons (the $2p_z$ electrons) on the oxygen atom in phenol becomes partly delocalised into the benzene ring π-bond ✓. This increases the electron density in the ring and makes it more susceptible to electrophilic attack ✓ and so has a lower activation energy, making the reaction faster ✓.

● Don't forget to make a comment about the activation energy. It is this that ʧers the rate.

uestion 6

is question is about isomers.

The correct name for the compound shown is:

A E-1-hydroxy-3-methylpent-2-ene **C** cis-1-hydroxy-3-methylpent-2-ene

B Z-1-hydroxy-3-methylpent-2-ene **D** $trans$-1-hydroxy-3-methylpent-2-ene (1 mark)

The compound $CH_3CH(OH)CH(CH_3)COOH$ has:

A no optical isomers   **C** three optical isomers

B two optical isomers   **D** four optical isomers (1 mark)

How many chiral centres has it got and is the molecule symmetrical?

A compound with the molecular formula C_4H_8O is thought to be one of:

$CH_3COCH_2CH_3$ $HCOCH_2CH_2CH_3$ $CH_2OHCH_2CH=CH_2$

X   Y   Z

ᵉ the following and the *Data Booklet* to identify the compound, explaining all the data.
ᵗ had no reaction with iodine in alkali.
ᵗ had IR peaks at $3400\,cm^{-1}$ and $3080\,cm^{-1}$ but none in the range $1740–1700\,cm^{-1}$.
ᵗs ^1H NMR spectrum had a singlet at $\delta = 2.8$ and a triplet at $\delta = 2.3$. (7 marks)

Work through the data logically. What groups do not react with iodoform?
ᵃt group causes an IR peak at $1740–1700\,cm^{-1}$? What causes a singlet and what
ᵘses a triplet?

(i) Which of the following compounds has an unsplit peak (a singlet) in its ^1H NMR spectrum?

    A CH_3CHO **C** $(CH_3)_2CHCHO$

    B $(CH_3)_3CH$ **D** $CH_3COCH_2CH_3$ (1 mark)

(ii) Explain why all peaks in ^{13}C NMR spectra are singlets. (1 mark)

Total: 11 marks

Questions & Answers

(a) A ✓

ℯ As there are four different groups around the C=C, the *cis/trans* method of naming geometric isomers cannot be used, so options C and D are incorrect. The group on the right-hand carbon of the C=C that has the higher priority is the C_2H_5 group; that with the higher priority on the left-hand carbon is the CH_2OH group. These are on opposite sides of the double bond, so the compound is an *E*-isomer.

(b) D ✓

ℯ The compound has two different chiral centres, as shown by the asterisks * on the carbon atoms in the formula:

$$H_3C-\overset{\overset{\displaystyle H}{|}}{\underset{\underset{\displaystyle OH}{|}}{C^*}}-\overset{\overset{\displaystyle H}{|}}{\underset{\underset{\displaystyle CH_3}{|}}{C^*}}-COOH$$

The four isomers are ++, −−, +− and −+. The last two are different because the four groups around the first chiral carbon atom are different from the four groups around the second chiral carbon atom.

(c) As it does not react with OH^-/I_2 it cannot be X ✓.

The IR peak at $3400\,cm^{-1}$ is caused by O–H stretching in alcohols ✓, that at $3080\,cm^{-1}$ by C–H in an alkene ✓. The absence of a peak between 1740 and $1700\,cm^{-1}$ shows that it does not contain a C=O group ✓. Thus the substance is neither X nor Y, so the substance is Z ✓.

The singlet at $\delta = 2.8$ in its ^1H NMR spectrum is due to the H of an OH group ✓, the triplet at $\delta = 2.3$ is due to the H in the CH_2 next to the CH_2OH, which is split into three by the two neighbouring H atoms ✓.

(d) (i) D ✓

ℯ A singlet can be due to the hydrogen of an OH group, or to the hydrogen atoms on a carbon atom that does not have an adjacent carbon atom with a hydrogen atom attached. Thus all alcohols and carboxylic acids have a singlet, as do methyl ketones, which have a CH_3CO group. The CH_3 hydrogen atoms in butanone (D) (on the left as written in the question) have a C=O group as their neighbour and so this peak will not be split. Ethanal (A) has two peaks, one split into four and the other into two. Methylpropane (B) has two peaks, one split into ten and the other into two. Methylpropanal (C) has three peaks, one split into seven and two split into two.

(ii) Because the chances of two ^{13}C atoms being next to each other in the molecule is effectively zero ✓.

Question 7

This question is about nitrogen compounds.

Nitrobenzene can be prepared from benzene by heating it under reflux at 50°C with a mixture of concentrated nitric and sulfuric acids.

Heating under reflux is used if the reaction is slow at room temperature and:

A   the product has a low boiling temperature

B   one of the reactants has a low boiling temperature

C   the reactants and products have similar boiling temperatures

D   the reactants and product form separate layers   (1 mark)

When amines and amides are dissolved in water, the pH of the solutions are:

A   amines pH > 7, amides pH = 7

B   amines pH < 7, amides pH = 7

C   amines pH = 7, amides pH > 7

D   amines pH = 7, amides pH < 7   (1 mark)

What are the pH ranges of acids and alkalis?

(i)   Write the equation for the reaction of 1-butylamine with chloroethane.   (1 mark)

(ii)   What would you observe if an amine such as ethylamine was added slowly until in excess to a solution containing copper(II) ions?   (2 marks)

A polyamide such as Kevlar®:

A   is water soluble because of hydrogen bonds between the –NH groups and water

B   is made from an amine such as ethylamine and a dicarboxylic acid such as benzene-1,3-dicarboxylic acid

C   is a solid with hydrogen bonding between strands

D   is biodegradable   (1 mark)

Remembering what Kevlar® is used for gets rid of two of the responses.

Total: 6 marks

Student answers

B ✓

Be careful here: it is the reactant, not the product, that must be prevented from boiling off, so option A is incorrect.

A ✓

Amines are bases, so their solutions will have a pH > 7. Amides are neutral.

(c) (i) $CH_3(CH_2)_3NH_2 + C_2H_5Cl \rightarrow CH_3(CH_2)_3N^+H_2(C_2H_5)Cl^- ✓$

allow $\rightarrow CH_3(CH_2)_3NH(C_2H_5) + HCl$

ⓔ The product is a secondary amine so will form a salt with the HCl.

> (ii) At first a pale blue precipitate ✓ (of copper(II) hydroxide) is formed. This dissolves in excess to form a dark blue solution ✓ (of $[Cu(C_2H_5NH_2)_4(H_2O)_2]^{2+}$).

(d) C ✓

ⓔ Kevlar® is a polyamide and so forms strong hydrogen bonds between strands with the $\delta+$ hydrogen of the –NH forming a hydrogen bond with the lone pair of the $\delta-$ oxygen atom in a C=O group in another strand. In addition, the benzene rings form van der Waals attractions between strands, making Kevlar® stronger than other polyamides. Option A is incorrect because the benzene rings are hydrophobic and the interstrand hydrogen bonding is so strong that Kevlar® is not water soluble. The monomers of a polyamide must have two functional groups. Ethylamine has only one functional group, so it cannot form polymers. Therefore, option B is incorrect. Kevlar® is not a natural polymer and so is not biodegradable. Therefore, option D is also incorrect.

■ Paper 3-type questions

Question 1

Two identical twins were supplied with some propan-1-ol. Twin A was told to prepare propanal and twin B to prepare a solution of propanoic acid.

(a) Describe the method that twin A should use. Include in your answer a labelled diagram of the apparatus. (6 marks)

ⓔ You must draw the diagram carefully, making sure that you get propanal and not propanoic acid. How would you heat the mixture (the word 'heat' is not sufficient)?

(b) Twin B used the following method:
- Mix propan-1-ol with excess sodium dichromate(VI) dissolved in dilute sulfuric acid in a round-bottomed flask and heat the mixture under reflux for 15 minutes.
- Allow it to cool and rearrange the apparatus for distillation.
- Heat and distil off the mixture of water and propanoic acid.

The proposed apparatus used for the distillation is shown in Figure 1. The diagram contains three errors. Identify these errors. (3 marks)

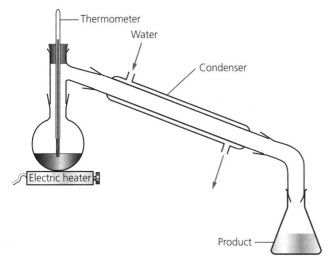

Figure 1

Both twins started with 5.0 g of propan-1-ol. Twin A made 3.1 g of propanal and twin B made 3.7 g of propanoic acid. Which twin had the higher percentage yield? Justify your answer.

(5 marks)

You will need to calculate molar masses.

Total: 14 marks

Student answers

Place the propan-1-ol and dilute sulfuric acid in a round-bottomed flask and heat it to a temperature above that of propanal and below that of propan-1-ol ✓. Then set it up as for distillation with addition. Slowly add a solution of sodium dichromate(VI) and distil off the propanal as it is formed ✓.

Apparatus marks: flask, with contents labelled ✓; with still head fitted with a dropping funnel labelled with sodium dichromate(VI) solution ✓; connected to a water condenser with water in at the bottom ✓. Collecting flask (immersed in iced water) with propanal labelled ✓.

The three errors are:

The thermometer should be opposite the entrance to the condenser ✓.

The water should go in at the lower end and out at the upper end of the condenser ✓.

The apparatus must have an opening at the collecting flask ✓.

molar masses/g mol⁻¹: propan-1-ol = 60, propanal = 58, and propanoic acid = 74 ✓

Theoretical yield of propanal = $5 \times \dfrac{58}{60}$ = 4.8 g ✓

Theoretical yield of propanoic acid = 6.2 g ✓

% yield of propanal = $\dfrac{3.1}{4.8} \times 100$ = 64.6%

% yield of propanoic acid = $\dfrac{3.7}{6.2} \times 100$ = 59.7% ✓

So twin A had the higher percentage yield ✓

Question 2

Aspirin can be prepared from 2-hydroxybenzoic acid by converting the phenolic group into an ethanoate ester, as shown in Figure 2.

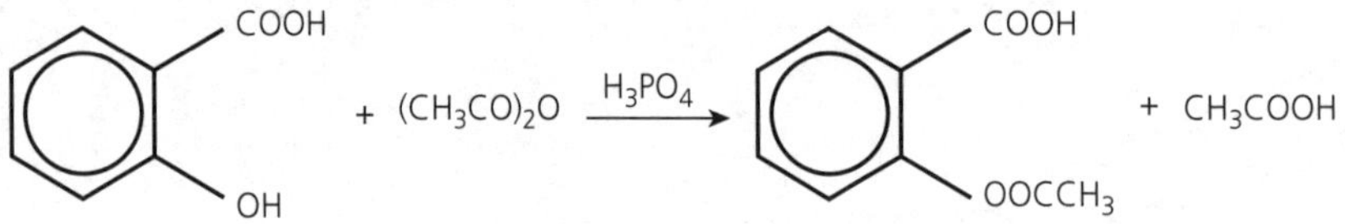

Figure 2

(a) Excess ethanoate anhydride was reacted with 2-hydroxybenzoic acid in the presence of phosphoric acid. When the reaction was complete, water was added, the mixture cooled and impure aspirin precipitated out. Describe how you would obtain a pure sample of aspirin from the reaction mixture. Note that aspirin is soluble in hot water but insoluble in cold water. (6 marks)

ⓔ In paper 3, the marking would be 4 marks for making six 'indicative' points and 2 marks for clearly expressed and logical writing of the procedure. This type of question will always require at least the number of points equal to the total mark, but you must write more than a series of brief bullet points.

(b) Explain how the purity of the aspirin could be checked. (2 marks)

(c) What would you observe if 2-hydroxybenzoic acid were added to:
   (i) bromine water
   (ii) solid sodium carbonate (2 marks)

Total: 10 marks

Student answers

(a) When cool, filter off the aspirin from the reaction mixture ✓.

Place it in a beaker/small conical flask and then dissolve it in the minimum of boiling water ✓.

While it is hot, filter off any undissolved solid (using a stemless funnel) ✓.

Allow the solution to cool ✓.

When it is at room temperature, filter off the solid under suction/using a Buchner funnel ✓.

Wash the solid with a little cold water and allow it to dry ✓.

All six indicative points = 4. Clearly explained = 2, reasonably well explained = 1, just a jumble of points = 0 to a total of 6 marks.

(b) Measure its melting temperature ✓.

Either: it should melt at the databook value (136°C) ±1°C

Or: it should melt at a sharp temperature ✓.

(c) (i) The red-brown bromine water will be decolorised ✓.

(ii) Bubbles of (CO_2) gas would be observed ✓.

Question 3

Ethanamide (CH_3CONH_2) can be prepared in the laboratory from ethanol. Outline how this could be carried out, stating the reagents and conditions needed and identifying all intermediate compounds formed. (6 marks)

It might help to work backwards. First: from what type of compound can you prepare an amide? Then: how is that type of compound prepared from an alcohol? Remember to give conditions such as heat under reflux and the names or formulae of the two intermediates.

An organic compound Z contains only carbon, hydrogen, nitrogen and oxygen. When 1.00 g of Z was burnt, 1.48 g of carbon dioxide, 0.71 g of water and 0.52 g of nitrogen dioxide (NO_2) were produced.

(i) Calculate the percentage by mass of the elements in Z. (2 marks)

Remember that the percentages of carbon, hydrogen, nitrogen and oxygen add to 100.

(ii) Use your answer to (i) to show that Z has the empirical formula $C_3H_7O_2N$. (2 marks)

It is helpful to set your calculation out in a table.

The molecular formula of compound Z is the same as its empirical formula. Compound Z melts at a high temperature, is soluble in water and reacts with both acids and alkalis at room temperature. When heated with ninhydrin, a dark colour is produced.

Which functional groups in organic chemistry react with acids and which with alkalis? What type of compound reacts with ninhydrin?

(i) Draw the structural formula of Z, which has a chiral centre. (2 marks)

You are told the molecular formula from the stem of (c) combined with the information in (b) (ii).

(ii) Explain why compound Z has a high melting temperature and why it is soluble in water. (4 marks)

High melting temperature means strong forces, but between what in this case? Solubility in water implies either hydrogen bonding with water molecules or being ionic.

(iii) Write ionic equations to show the reactions of compound Z with $H^+(aq)$ ions and with $OH^-(aq)$ ions. (2 marks)

Remember that ionic equations must balance for charge.

Total: 18 marks

Questions & Answers

(a) Step 1: $C_2H_5OH \rightarrow CH_3COOH$ ✓ (ethanoic acid)

Reagents: acidified potassium dichromate ✓ ($H^+/Cr_2O_7^{2-}$)

Conditions: heat under reflux ✓

Step 2: $CH_3COOH \rightarrow CH_3COCl$ ✓ (ethanoyl chloride)

Reagent: phosphorus pentachloride ✓ (PCl_5)

Step 3: $CH_3COCl \rightarrow CH_3CONH_2$

Reagent: ammonia solution ✓ (NH_3)

ⓔ Alternative reagents for step 2 are PCl_3 or $SOCl_2$.

(b) (i) mass of carbon = $1.48 \times \dfrac{12.0}{44.0} = 0.404\,g$ % carbon = 40.4%

mass of hydrogen = $0.71 \times \dfrac{2.0}{18.0} = 0.0789\,g$ % hydrogen = 7.89%

mass of nitrogen = $0.52 \times \dfrac{14.0}{46.0} = 0.158\,g$ % nitrogen = 15.8%

% oxygen = 100 − 40.4 − 7.89 − 15.8 = 35.91%

All four correct ✓✓; three correct ✓

ⓔ Note that to obtain the mass of hydrogen, the mass of water has to be multiplied by 2, because there are two hydrogen atoms in each water molecule. In this example the percentage is 100 times the mass because the initial mass of Z was 1.00 g.

(ii)

Element	%	Divide by A_r ✓	Divide by smallest ✓
Carbon	40.4	40.4/12.0 = 3.37	3.37/1.13 = 3.0
Hydrogen	7.89	7.89/1.0 = 7.89	7.89/1.13 = 7.0
Nitrogen	15.8	15.8/14.0 = 1.13	1.13/1.13 = 1.0
Oxygen	35.91	35.91/16.0 = 2.24	2.24/1.13 = 2.0

This is consistent with an empirical formula of $C_3H_7O_2N$.

(c) (i) The formula is:

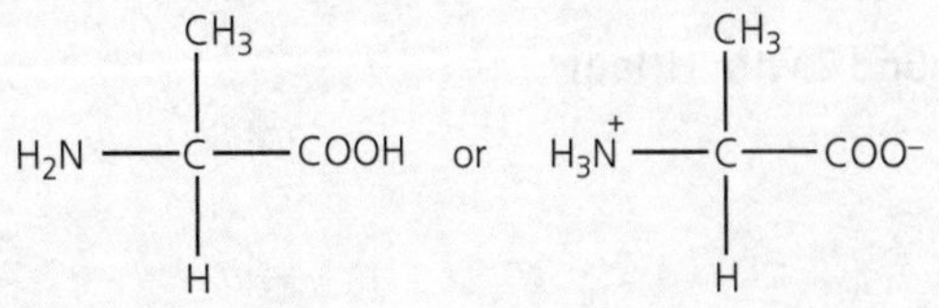

ⓔ The information in the question indicates that Z is an amino acid. Ninhydrin reacts with amino acids to form a dark coloration. This is how the positions of different amino acids are identified in thin-layer chromatography. Either the molecular formula or the formula of the zwitterion is acceptable. 1 mark is for identifying the two functional groups and the other for the rest of the molecule.

(ii) Compound Z has a high melting temperature because it forms a zwitterion, $^+H_3NCH(CH_3)COO^-$ ✓. There are strong ionic forces of attraction between different zwitterions. This means that a lot of energy is required to separate zwitterions and therefore the melting temperature is high ✓.

You must make it clear that the ion/ion forces are between *different* zwitterions.

It is soluble in water because hydrogen bonds form between the COO^- groups and the $\delta+$ hydrogen atoms in water ✓ and between the H_3N^+ groups and the lone pairs of electrons on the $\delta-$ oxygen atoms in water ✓.

1 mark would be allowed for explaining how the molecular form of Z would m hydrogen bonds with the water.

(iii) Either: $H_2NCH(CH_3)COOH + H^+ \rightarrow {}^+H_3NCH(CH_3)COOH$
or: $^+H_3NCH(CH_3)COO^- + H^+ \rightarrow {}^+H_3NCH(CH_3)COOH$ ✓
and either: $H_2NCH(CH_3)COOH + OH^- \rightarrow H_2NCH(CH_3)COO^- + H_2O$
or: $^+H_3NCH(CH_3)COO^- + OH^- \rightarrow H_2NCH(CH_3)COO^- + H_2O$ ✓

uestion 4

(i) Compound Q contains carbon, hydrogen and oxygen. When 1.52 g of Q was burnt in excess oxygen, 2.64 g of carbon dioxide and 1.44 g of water were produced. Calculate its empirical formula. (3 marks)

(ii) Its mass spectrum showed a molecular ion of $m/z = 76$. State its molecular formula. (1 mark)

The ^1H NMR spectrum of compound Q is shown in Figure 2.

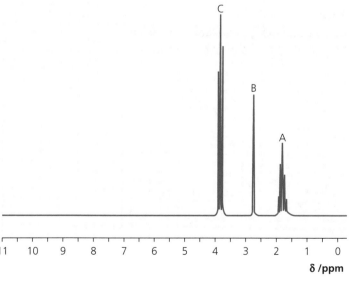

Figure 2

(i) Explain why this is the spectrum of propan-1,3-diol and not that of propan-1,2-diol. Include in your answer the number of peaks and their splitting patterns. (3 marks)

(ii) Explain why the set of peaks marked A is split into five and that marked B is a singlet. (2 marks)

(c) The infrared spectrum of propan-1,3-diol is shown in Figure 3.

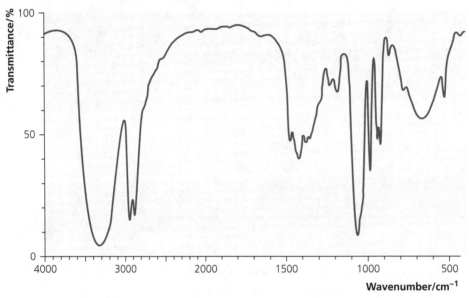

Figure 3

(i) Identify the groups that cause the peaks at $2945\,cm^{-1}$ and at $3350\,cm^{-1}$. (2 marks)

(ii) Why is the peak at $3350\,cm^{-1}$ broad? (2 marks)

(d) Propan-1,3-diol is heated under reflux with excess potassium dichromate(VI) in dilute sulfuric acid.

(i) Write the skeletal formula of the organic product. (1 mark)

(ii) The infrared spectrum of this product has an extra peak. State the group causing this peak and give an estimate of the wave number of this peak. (2 marks)

Total: 15 marks

Student answers

(a) (i) mass C $= 2.64 \times \dfrac{12}{44} = 0.72\,g$ which is $0.72 \div 12 = 0.06\,mol$ ✓

mass H $= 1.44 \times \dfrac{2}{18} = 0.16\,g$ which is $0.16\,mol$

mass O $= 1.52 - (0.72 + 0.16) = 0.64\,g$ which is $0.64/16 = 0.04\,mol$ ✓

ratio C:H:O $= 0.06:0.16:0.04 = 3:8:2$, so the empirical formula is $C_3H_8O_2$ ✓

(ii) The empirical mass is 76, so the molecular formula is $C_3H_8O_2$ ✓

(i) Propan-1,3-diol will have three lines (as the molecule is symmetrical) whereas propan-1,2-diol will have five lines✓. The lines in propan-1,3-diol will be split into 5, 3 and a singlet✓ whereas those of propan-1,2-diol will be two doublets, one split into six and two singlets✓ (although their δ values will be very similar).

(ii) The peak A is split into five, caused by there being four hydrogen atoms attached to adjacent carbon atom(s)✓ and that at B is caused by the hydrogen atom of the OH group.

(i) The peak at 2945 cm^{-1} is caused by a C–H stretch ✓ and that at 3350 cm^{-1} by an O–H stretch✓.

(ii) It is broad because of hydrogen bonding✓.

(i)

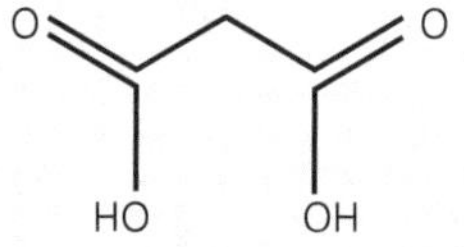

(ii) It will be caused by the C=O group✓ at a value between 1725 and 1700 cm^{-1} ✓.

Question 5

A dipeptide is thought to be made from glycine and alanine. It was hydrolysed to produce a solution of the two amino acids. You are also given samples of alanine and glycine.

Describe how the amino acids in the mixture could be separated and identified. (4 marks)

Explain why alanine, $CH_3CH(NH_2)COOH$:

(i) has a melting temperature in excess of 100°C (3 marks)

(ii) is soluble in water (2 marks)

You must describe the forces involved in these processes.

Write ionic equations for the reactions of alanine with:

(i) aqueous acid (1 mark)

(ii) aqueous alkali (1 mark)

(i) State the reagents and conditions for the three-step synthesis of $CH_3CH(NH_2)COOH$ from 3-hydroxypropene, $CH_2=CHCH_2OH$. (5 marks)

Think how an NH_2 group can be introduced into a molecule and how a COOH group can be made from an alcohol.

(ii) State one way that the physical properties of the product in (i) differs from that of natural alanine. (1 mark)

Total: 17 marks

Questions & Answers

(a) A spot of the solution is placed on a thin-layer chromatography plate with spots of alanine and glycine at the same height ✓. It is then placed in a tank with eluent so that the level of the liquid eluent is just below the line of the spots ✓. The plate is left until the eluent has risen near to the top of the plate. The plate is then dried and sprayed with ninhydrin ✓. One spot will be shown level with that of the alanine and one level with that of glycine ✓.

(b) (i) Alanine exists as a zwitterion, $CH_3CH(NH_3^+)COO^-$ ✓. The ion–ion forces between different zwitterions ✓ are strong and so a high temperature is required to supply the energy necessary to separate the zwitterions ✓.

(ii) The N^+ form strong bonds with the $\delta-$ oxygen in water✓ and the O^- forms strong bonds with the $\delta+$ hydrogen in water ✓, making alanine water soluble.

(c) (i) $CH_3CH(NH_3^+)COO^- + H^+ \rightarrow CH_3CH(NH_3^+)COOH$ ✓

or: $CH_3CH(NH_2)COOH + H^+ \rightarrow CH_3CH(NH_3^+)COOH$

(ii) $CH_3CH(NH_3^+)COO^- + OH^- \rightarrow CH_3CH(NH_2)COO^- + H_2O$ ✓

or: $CH_3CH(NH_2)COOH + OH^- \rightarrow CH_3CH(NH_2)COO^- + H_2O$

(d) (i) Step 1: $CH_2CH=CH_2OH \rightarrow CH_3CHClCH_2OH$ ✓ bubble in hydrogen chloride ✓

Step 2: $CH_3CHClCH_2OH \rightarrow CH_3CH(NH_2)CH_2OH$ ✓ add concentrated ammonia ✓

Step 3: $CH_3CH(NH_2)CH_2OH \rightarrow CH_3CH(NH_2)COOH$, heat under reflux with potassium dichromate in acid solution✓.

(ii) It will have no effect on the plane of polarisation of plane polarised light (as it will be the racemic mixture) whereas natural alanine will rotate the plane of polarisation ✓.

Question 6

Read the passage carefully and answer the questions that follow.

Electromagnetic radiation

Electromagnetic radiation is important in chemistry in a number of ways.

Radio waves of frequency around 100–400 MHz are used in proton NMR spectroscopy. The substance under investigation is placed in a very strong magnetic field. This causes the energy levels of spinning hydrogen nuclei to split into two. The peaks in the NMR spectrum measure the difference between these two levels, which varies according to the exact environment of the hydrogen nuclei. The peaks are also split by the hydrogen atoms on neighbouring carbon atoms; the splitting follows the (n + 1) rule. Microwaves have a higher frequency. This radiation is only absorbed by polar molecules. The molecules then rotate faster gaining rotational energy. On collision with another molecule, this energy is converted into kinetic energy, in the form of heat. Use of this is made in the pharmaceutical industry where microwaves are used to heat up reaction mixtures.

frared spectroscopy uses radiation of a higher frequency still. It detects the
esence of functional groups and can be used to identify unknown compounds.
hen infrared radiation is absorbed by a molecule, one or more of its bonds
nd or stretch. Absorption only takes place if the stretching or bending causes a
ange in dipole moment.

✓ light is the highest frequency radiation normally used in chemistry. It causes
nds in a molecule to break homolytically, forming free radicals. When a mixture
 chlorine and methane is exposed to a short pulse of UV radiation, chlorine
dicals are produced. These then take part in a chain reaction, the first step being
e removal of a hydrogen atom from methane to form hydrogen chloride and a
ethyl radical.

Place the four types of electromagnetic radiation discussed in the passage in
order of decreasing frequency. (1 mark)

Explain why the air inside a microwave oven does not heat up when a cup of
water is being heated in the oven. (1 mark)

(i) State and explain the difference between the high-resolution proton NMR
spectra of ethanol (CH_3CH_2OH) and ethanamide (CH_3CONH_2). (5 marks)

You need to state and explain the number of peaks in each spectrum and their
itting patterns.

(ii) The infrared spectra of ethanol and ethanamide compounds are shown
as spectrum X and spectrum Y in Figure 3. Identify the spectrum of
ethanamide, giving reasons. (You should consult the *Data Booklet*.) (3 marks)

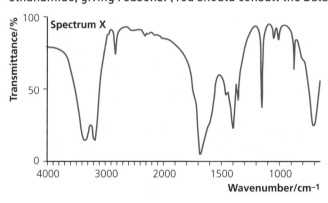

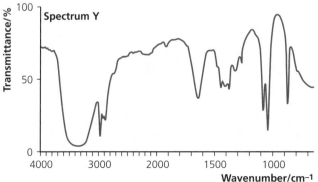

Figure 3

ℯ The only peaks that need to be considered are those over $1500\,cm^{-1}$. Identify the frequency and the bonds that cause the different main peaks in each spectrum.

(d) When chlorine gas is bubbled into liquid methylbenzene ($C_6H_5CH_3$) in the presence of UV radiation, a free radical substitution reaction takes place. One of the products is chloromethylbenzene ($C_6H_5CH_2Cl$). The mechanism of this reaction is similar to that between methane and chlorine.

   (i) Write the mechanism of the reaction between chlorine gas and methylbenzene, showing the initiation step and two propagation steps.   (3 marks)

ℯ Make sure that radicals are identified by a dot.

   (ii) Some 1,2-diphenylethane ($C_6H_5CH_2CH_2C_6H_5$) is formed in this reaction. Explain how this is evidence for the mechanism that you have written in (i).   (2 marks)

   (iii) Suggest how a pure dry sample of chloromethylbenzene could be extracted from the reaction mixture.   (5 marks)

Compound	Density/$g\,cm^{-3}$	Boiling temperature/°C	Solubility in water
Chloromethylbenzene	1.1	179	Insoluble
Methylbenzene	0.87	111	Insoluble

Total: 20 marks

ℯ You must remember that the final reaction mixture will contain both products of the wanted reaction plus some unreacted starting reagents. The hydrogen chloride formed in the reaction will be dissolved in the chloromethylbenzene. How can it be removed and how can a dry mixture of organic substances be obtained? How can the desired product be obtained from this mixture?

Student answers

(a) UV > IR > microwaves > radio waves ✓

(b) The air consists mostly of oxygen and nitrogen. These are non-polar molecules and so do not absorb microwaves ✓.

(c) (i) Ethanol has the formula CH_3CH_2OH and has hydrogen nuclei in three chemical environments. Its ^1H NMR spectrum has three different peaks ✓. The peak due to the hydrogen atoms in the CH_3 group is split into three as there are two H atoms on the neighbouring carbon atom ✓ and those due to the hydrogen atoms in the CH_2 group are split into four by the three H atoms of the neighbouring CH_3 ✓. The peak due to the hydrogen in an OH group is never split ✓.
Ethanamide has the formula CH_3CONH_2 and has hydrogen nuclei in two chemical environments. Its spectrum has two peaks and neither of the peaks is split ✓.

ℯ Don't forget to describe for both molecules the number of peaks and the extent of splitting due to spin coupling ($n + 1$) rule. The peaks in the ethanamide spectrum are not split, because there is no hydrogen atom on the carbon atoms adjacent to the CH_3 and NH_2 groups.

(ii) Both spectra have a broad peak at around $3350\,\text{cm}^{-1}$. These are caused by the O–H bond in ethanol and the N–H bond in ethanamide ✓. Spectrum X has a strong peak at $1680\,\text{cm}^{-1}$. This is typical of the C=O bond in amides (1700–$1630\,\text{cm}^{-1}$) ✓, so spectrum X is that of ethanamide and spectrum Y is that of ethanol ✓.

4) (i) Initiation step: $\text{Cl}_2 \xrightarrow{\text{UV}} 2\text{Cl}\bullet$ ✓

Propagation steps:

$\text{Cl}\bullet + \text{C}_6\text{H}_5\text{CH}_3 \rightarrow \text{C}_6\text{H}_5\text{CH}_2\bullet + \text{HCl}$ ✓

$\text{C}_6\text{H}_5\text{CH}_2\bullet + \text{Cl}_2 \rightarrow \text{C}_6\text{H}_5\text{CH}_2\text{Cl} + \text{Cl}\bullet$ ✓

(ii) The 1,2-diphenylethane is formed by two $\text{C}_6\text{H}_5\text{CH}_2\bullet$ radicals joining together in a chain termination step ✓.

$\text{C}_6\text{H}_5\text{CH}_2\bullet + \text{C}_6\text{H}_5\text{CH}_2\bullet \rightarrow \text{C}_6\text{H}_5\text{CH}_2\text{CH}_2\text{C}_6\text{H}_5$ ✓

(iii) The reaction mixture will contain some hydrogen chloride, unreacted methylbenzene and some polysubstituted methylbenzene. The following procedure should be used to obtain pure chloromethylbenzene:

1 Wash the reaction mixture with sodium carbonate (or sodium hydrogencarbonate) solution and then with water ✓ in a separating funnel, collecting the lower organic layer ✓ each time.

This removes hydrogen chloride, which is a reaction product impurity.

2 Dry the organic layer with anhydrous calcium chloride ✓.

3 Filter off the calcium chloride and distil ✓ the clear dried organic layer. Discard any unreacted methylbenzene that distils over at around 111°C and collect the fraction that boils over between 177°C and 181°C ✓.

The polysubstituted methylbenzenes have a higher boiling temperature and left behind in the distillation flask.

Knowledge check answers

1 $[OH^-]$ decreased from $0.00020\,mol\,dm^{-3}$ to zero in 26 s so initial rate $= (0.00020 - 0)/26 = 7.7 \times 10^{-6}\,mol\,dm^{-3}\,s^{-1}$

2 $k = rate/[A][B]$

units of $k = (mol\,dm^{-3}\,s^{-1})/(mol\,dm^{-3} \times mol\,dm^{-3})$
$= mol^{-1}\,dm^{+3}\,s^{-1}$

3 rate $= k[A] \times [C]$

4 Because sodium hydroxide contains OH^- ions. OH^- ions are a good catalyst for the reaction between propanone and iodine. Iodoform, CHI_3, will be formed rapidly (see page 28).

5 Hydrolysis of halogenoalkanes involves the breaking of the carbon–halogen bond. As the C–I bond is weaker than the C–Cl bond, less energy is required to break it. Thus the activation energy will be less causing the rate constant, k, to be bigger and hence the reaction faster.

6 An electrophile is a species that attacks electron-rich sites by accepting a pair of electrons and forming a covalent bond.

A nucleophile is a species with a lone pair of electrons which it uses to form a bond with a $\delta+$ atom.

7 cis (or Z-) but-2-ene, trans (or E-) but-2-ene, but-1-ene, and methylpropene

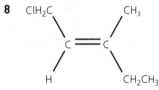

Or: $CH_2=CHCH_2CH_3$ $CH_2=C(CH_3)CH_3$

8

9 Because the product is not chiral. (There are not four different groups around the carbon atom.)

10 It will be positive as a gas is being produced from a solution and a solid. Gases are more random and have a higher entropy.

11 The bond energy calculation uses *average* bond enthalpies, so is less accurate than that calculated from enthalpy of hydrogenation.

12 The methyl group is an electron-releasing (electron-pushing) group and so increases the electron density in the ring. This makes the attack by electrophiles have a lower activation energy.

13 Ethylamine and propane have the same number of electrons and hence similar strength London forces. However, ethylamine also forms strong intermolecular hydrogen bonds. Thus more energy is required to separate ethylamine molecules and so it has a higher boiling temperature.

14 The volatile ethylamine reacts with acids to form an involatile ionic salt. Thus the smell disappears. When alkali is then added, the volatile ethylamine is liberated and so the smell returns. The equations with HCl as the acid are:

$C_2H_5NH_2 + HCl \rightarrow C_2H_5NH_3^+Cl^-$

$C_2H_5NH_3^+Cl^- + OH^- \rightarrow C_2H_5NH_2 + H_2O$

15

16 $^+NH_3CH_2COO^- + OH^- \rightarrow NH_2CH_2COO^- + H_2O$

17 NH_2—CH—CONH—CH_2COOH

and

NH_2—CH_2—CONH—CH—COOH

18 $CH_2OHCH=CHCHO$ (or $CH_3C(OH)=CHCHO$ or $CH_2C=CHCH(OH)CHO$ or $CH_2=C(CHO)CH_2OH$). It m have a C=C, an OH and an aldehyde group.

19 Step 1 Add a mixture of concentrated nitric and sulfuric acids at 50°C.

Step 2 Warm the nitrobenzene formed in step 1 w tin and concentrated hydrochloric acid then add alkali and steam distil off the phenylamine forme

Step 3 Add ethanoyl chloride, CH_3COOCl.

20 (a) The distillation apparatus (Figure 54).

(b) The heat under reflux apparatus (Figure 52).

21 The H of the CH_3 will be a triplet at $\delta = 0$ to 1.8 (actually 1.1), the H of the CH_2 will be split into fou $\delta = 1.8$ to 2.8 (actually 2.5) and the H of CHO will b triplet at $\delta = 9$ to 10 (actually 9.8).

22 The chances of two ^{13}C atoms next to each other in the carbon chain are negligible, so no splitting occurs.

23 The IR peak at $3410\,cm^{-1}$ is due to an alcoholic OF group not that in a carboxylic acid. The peak at $1710\,cm^{-1}$ is due to C=O in a ketone or carboxylic acid, but the substance is not an acid because of OH value. The peak in the mass spectrum at m/z 57 is caused by the loss of CH_2OH (88 – 31) and th at 43 by CH_3CO^+. The only substance that fits the data is:

Index

Periodic table

The periodic table

Key:

Relative atomic mass
Atomic symbol
name
Atomic (proton) number

Group

Period	1	2													3	4	5	6	7	0
1	1.0 H hydrogen 1																			4.0 He helium 2
2	6.9 Li lithium 3	9.0 Be beryllium 4													10.8 B boron 5	12.0 C carbon 6	14.0 N nitrogen 7	16.0 O oxygen 8	19.0 F fluorine 9	20.2 Ne neon 10
3	23.0 Na sodium 11	24.3 Mg magnesium 12													27.0 Al aluminium 13	28.1 Si silicon 14	31.0 P phosphorus 15	32.1 S sulfur 16	35.5 Cl chlorine 17	39.9 Ar argon 18
4	39.1 K potassium 19	40.1 Ca calcium 20	45.0 Sc scandium 21	47.9 Ti titanium 22	50.9 V vanadium 23	52.0 Cr chromium 24	54.9 Mn manganese 25	55.8 Fe iron 26	58.9 Co cobalt 27	58.7 Ni nickel 28	63.5 Cu copper 29	65.4 Zn zinc 30			69.7 Ga gallium 31	72.6 Ge germanium 32	74.9 As arsenic 33	79.0 Se selenium 34	79.9 Br bromine 35	83.8 Kr krypton 36
5	85.5 Rb rubidium 37	87.6 Sr strontium 38	88.9 Y yttrium 39	91.2 Zr zirconium 40	92.9 Nb niobium 41	95.9 Mo molybdenum 42	[98] Tc technetium 43	101.1 Ru ruthenium 44	102.9 Rh rhodium 45	106.4 Pd palladium 46	107.9 Ag silver 47	112.4 Cd cadmium 48			114.8 In indium 49	118.7 Sn tin 50	121.8 Sb antimony 51	127.6 Te tellurium 52	126.9 I iodine 53	131.3 Xe xenon 54
6	132.9 Cs caesium 55	137.3 Ba barium 56	138.9 La lanthanum 57	178.5 Hf hafnium 72	180.9 Ta tantalum 73	183.8 W tungsten 74	186.2 Re rhenium 75	190.2 Os osmium 76	192.2 Ir iridium 77	195.1 Pt platinum 78	197.0 Au gold 79	200.6 Hg mercury 80			204.4 Tl thallium 81	207.2 Pb lead 82	209.0 Bi bismuth 83	[209] Po polonium 84	[210] At astatine 85	[222] Rn radon 86
7	[223] Fr francium 87	[226] Ra radium 88	[227] Ac actinium 89	[261] Rf rutherfordium 104	[262] Db dubnium 105	[266] Sg seaborgium 106	[264] Bh bohrium 107	[277] Hs hassium 108	[268] Mt meitnerium 109	[271] Ds darmstadtium 110	[272] Rg roentgenium 111									

Elements with atomic numbers 112–116 have been reported but not fully authenticated

140.1 Ce cerium 58	140.9 Pr praseodymium 59	144.2 Nd neodymium 60	144.9 Pm promethium 61	150.4 Sm samarium 62	152.0 Eu europium 63	157.2 Gd gadolinium 64	158.9 Tb terbium 65	162.5 Dy dysprosium 66	164.9 Ho holmium 67	167.3 Er erbium 68	168.9 Tm thulium 69	173.0 Yb ytterbium 70	175.0 Lu lutetium 71
232 Th thorium 90	[231] Pa protactinium 91	238.1 U uranium 92	[237] Np neptunium 93	[242] Pu plutonium 94	[243] Am americium 95	[247] Cm curium 96	[245] Bk berkelium 97	[251] Cf californium 98	[254] Es einsteinium 99	[253] Fm fermium 100	[256] Md mendelevium 101	[254] No nobelium 102	[257] Lr lawrencium 103